DIFFRACTION THEORY AND ANTENNAS

ELLIS HORWOOD SERIES IN ELECTRICAL AND ELECTRONIC ENGINEERING

Series Editor:
P. S. Brandon, *Professor of Electrical Engineering*
University of Cambridge

APPLIED CIRCUIT THEORY: Matrix and Computer Methods
P. R. ADBY, University of London King's College.

NOISE IN SOLID STATE DEVICES
D. A. BELL, University of Hull and M. J. BUCKINGHAM, Royal Aircraft Establishment, Farnborough, Hampshire.

DIFFRACTION THEORY AND ANTENNAS
R. H. CLARKE and JOHN BROWN, Imperial College of Science and Technology, University of London.

INTEGRATED CIRCUIT TECHNOLOGY OF COMMUNICATION-BASED COMPUTERS
C. MOIR, Ministry of Defence, Royal Signals and Radar Establishment, Worcestershire.

CONTROL SYSTEMS WITH TIME DELAYS
A. Olbrot, Institute of Automatic Control, Technical University of Warsaw.

HANDBOOK OF RECTIFIER CIRCUITS
G. J. SCOLES, English Electric Valve Company, Chelmsford, Essex.

PRINCIPLES OF COMPUTER COMMUNICATION NETWORK DESIGN
J. SEIDLER, Institute of Information Science, Technical University of Gdansk, Poland.

DIFFRACTION THEORY AND ANTENNAS

R. H. CLARKE and
JOHN BROWN
Imperial College of Science and Technology
University of London

ELLIS HORWOOD LIMITED
Publishers · Chichester

Halsted Press: a division of
JOHN WILEY & SONS
New York · Chichester · Brisbane · Toronto

First published in 1980 by

ELLIS HORWOOD LIMITED

Market Cross House, Cooper Street, Chichester, West Sussex, PO19 1EB, England

The publisher's colophon is reproduced from James Gillison's drawing of the ancient Market Cross, Chichester.

Distributors:

Australia, New Zealand, South-east Asia:
Jacaranda-Wiley Ltd., Jacaranda Press,
JOHN WILEY & SONS INC.,
G.P.O. Box 859, Brisbane, Queensland 40001, Australia.

Canada:
JOHN WILEY & SONS CANADA LIMITED
22 Worcester Road, Rexdale, Ontario, Canada.

Europe, Africa:
JOHN WILEY & SONS LIMITED
Baffins Lane, Chichester, West Sussex, England.

North and South America and the rest of the world:
Halsted Press: a division of
JOHN WILEY & SONS
605 Third Avenue, New York, N.Y. 10016, U.S.A.

British Library Cataloguing in Publication Data
Clarke, Richard Henry
Diffraction theory and antennas. –
(Ellis Horwood series in electrical and electronic engineering).
1. Radio waves – Diffraction
2. Radio – Antennas
I. Title II. Brown, John, b. 1923
621.3841'1 TK6553 80–40388
ISBN 0–85312–182–6 (Ellis Horwood Ltd., Publishers – Library Edition)
ISBN 0–470–27003–9 (Halsted Press)

Typeset in Press Roman by Ellis Horwood Ltd.
Printed in Great Britain by Fakenham Press

Table of Contents

Authors' Preface

Diffraction theory is generally thought to be a difficult subject – one that is best left to a few experts. It is our belief, however, that the angular plane-wave spectrum approach makes the subject accessible to more than just a favoured few. Indeed, the student can go quite a long way armed with a rudimentary knowledge of plane waves and a familiarity with the description of signals in terms of their Fourier-transform spectra. It then comes as a pleasant surprise to find that the theory is not only more comprehensible but is also more precise than conventional theories of diffraction. The theory is at the same time useful and relatively easy to apply, as we show for several different microwave antennas.

We gladly acknowledge our particular debt to H. G. Booker and J. A. Ratcliffe who, with their students, have been largely responsible for promulgating the angular plane-wave spectrum approach. The method of presentation we have adopted has been greatly influenced by them. The description of two-dimensional diffracted fields in Chapters 2 and 5 is essentially an elaboration of Ratcliffe's (1956) review article; and the transmission-line approach to plane-wave and guided-wave propagation in Appendix A stems from Booker (1947). We are also very grateful to our own students and colleagues, both here and elsewhere, for providing us with the incentive to write this book; and to R. Puddy for the drawings and J. Ronen for some of the calculations.

The order of presentation reflects an attitude we commonly adopt in teaching, of arguing from the particular to the general. We have therefore deliberately placed Maxwell's equations at the end of the book in Appendix B, although we do refer to them occasionally in earlier parts of the text. Maxwell's equations are things of great power and beauty – they are what attracted many of us to electrical engineering in the first place – but if they are unfamiliar they are just a hindrance.

The policy we have adopted in quoting the work of others is to do so only if it arises naturally in the course of the exposition. It has not been our intention to identify the originator of all the ideas mentioned, or to trace parallel developments. We felt this would have made the work bottom-heavy with references, when the objective is to introduce the reader to new ideas.

R. H. CLARKE
JOHN BROWN
Imperial College, London
May, 1980

CHAPTER 1

Introduction

I have finally judged that it was better worth while to publish this writing, such as it is, than let it run the risk, by waiting longer, of remaining lost.

Christiaan Huygens
Treatise on Light,†
8 January 1690

Our understanding of the phenomenon of diffraction really begins with Huygens' famous construction, now known as Huygens' principle. This states that each point on a propagating wavefront can be considered as a secondary source radiating a spherical wave. The transverse polarization of the radiated wave was not apparent to Huygens; nor was the principle of wave interference, which had to wait more than a hundred years for Thomas Young to discover it. Augustin Fresnel then combined the ideas of Huygens and Young in order to describe the nature of the light diffracted by various apertures, such as a knife edge, an opaque strip, and a parallel-sided opening. This last example provides a useful introduction to the phenomenon of diffraction and its application to aperture antennas.

1.1 AN ELEMENTARY EXAMPLE OF DIFFRACTION

Consider a plane electromagnetic wave incident normally on a parallel-sided slit cut in a thin perfectly conducting plane, as in Fig. 1.1. For this two-dimensional situation a typical element of width dx in the aperture formed by the slit can be supposed, by an obvious modification of Huygens' principle, to radiate a cylindrical wave into the medium to the right of the aperture plane. It is reasonable to assume that the strength of the secondary source is proportional to the magnitude E_0 of the field incident on the aperture from the left, and to the elemental width dx. Then, without bothering with any refinements at this stage, the contribution of the element E_0dx to the field at a point P a distance r' away is

$$\mathrm{d}E_\mathrm{p} = CE_0 \mathrm{d}x \frac{1}{\sqrt{r'}} \exp(-jkr') \quad , \tag{1.1}$$

†Traité de la lumière. Translation by S. P. Thompson, Macmillan 1912, Dover 1962.

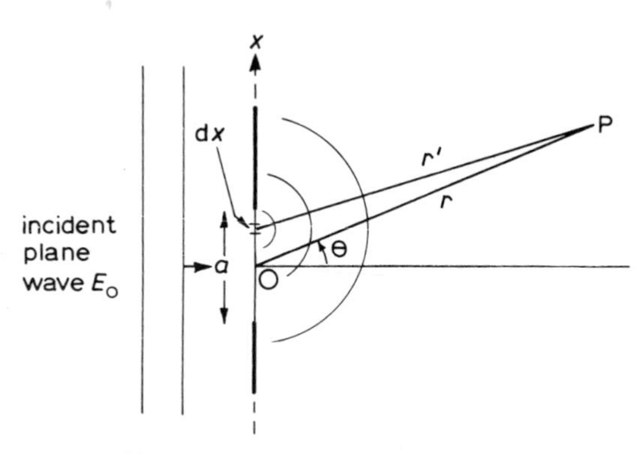

Fig. 1.1 – Plane wave diffracted by a parallel-sided slit cut in a perfectly conducting plane. Huygens representation.

where C is a constant and k is the phase retardation per unit distance suffered by a wave travelling in the medium to the right of the aperture plane at the frequency of the incident field. If the distance r of the point P from the centre O of the aperture is very large, in comparison with the aperture width a, it is legitimate to replace r' by r under the square root, and r' by $r - x \sin \theta$ in the phase term of equation (1.1). (The direction OP makes and angle θ to the axis normal to the aperture). Then, integrating over the width of the aperture, the total diffracted field at P is

$$E_p = \frac{CE_o}{\sqrt{r}} \exp(-jkr) \int_{-a/2}^{a/2} \exp(jkx \sin \theta) \, dx \tag{1.2}$$

so that

$$E_p = \frac{CE_o a}{\sqrt{r}} \exp(-jkr) \frac{\sin\left(\dfrac{\pi a \sin \theta}{\lambda}\right)}{\left(\dfrac{\pi a \sin \theta}{\lambda}\right)} \tag{1.3}$$

Thus Huygens' principle embodied in equation (1.1) gives the field at P as a superposition of cylindrical waves from all parts of the aperture (equation (1.2)) which produces the interference pattern of equation (1.3).

This equation reveals that the field at the distant point P diffracted by the aperture at O is basically a cylindrical wave centred on O, with an angular dependence of the form $(\sin \psi)/\psi$ where $\psi = \pi a \sin \theta/\lambda$. The angular dependence has a characteristic lobe structure, with the main lobe having its maximum in the direction $\theta = 0$, and with zeros at angles given by $\sin \theta = \pm \lambda/a, \pm 2\lambda/a$, and so on.

The uniformly illuminated slit is a useful elementary model of an aperture antenna, which is a class of antennas that is widely used at microwave frequencies for communication and radar surveillance. Thus it is immediately apparent from equation (1.3) that the level of the radiated field depends on the aperture width, increasing as a increases. At the same time the angular width of the main lobe is correspondingly reduced. In antenna parlance (see section 1.4), as the aperture width is increased, the gain of the antenna is increased and its beam-width is reduced.

1.2 DIFFRACTION THEORY USING PLANE WAVES

Developments in diffraction theory have been dominated by the ideas of Huygens and Fresnel. So, although considerable refinements have been achieved (see Stratton (1941), for example) in Kirchhoff's scalar theory of diffraction and then in Stratton and Chu's vector theory, the underlying principle of spherical waves radiated by known fields has remained. This is understandable in view of the simplicity and strength of Huygens' original idea. But it has meant that the student of diffraction has been presented with some difficult mathematical hurdles to clear before he can obtain some familarity with the phenomenon of diffraction.

The difficulty seems to arise from the fact that whereas spherical waves are a natural physical entity, they are rather clumsy from a mathematical point of view. In contrast, the plane wave, which is a simple mathematical entity, can never exist as such in the real physical world. However, it has long been known that naturally occurring fields can be *represented* by the superposition of either a discrete set or a continuum of plane waves travelling in different directions. Perhaps the simplest and best known example of this is the field in a rectangular waveguide, which is given precisely by the superposition of two plane waves travelling in directions equally inclined to the waveguide axis (see Appendix A.10). In general such a set of plane waves is known as an angular spectrum.

In this book we shall be using the angular plane-wave spectrum concept to develop the theory of diffraction. This approach is not only relatively simple mathematically, as already indicated, but it is also fundamentally more precise than older theories of diffraction. This is because the approximations that must inevitably be made occur at a later stage of the analysis in the case of the plane-wave approach.

One conceptual difficulty concerning the physical plausibility of the angular spectrum method must be dealt with right away. It is that, since plane waves are of infinite lateral extent, it may seem strange at first sight that they can be used to represent realistic fields. But the situation is no different, in principle, from the representation of an arbitrary time waveform $f(t)$ by the superposition of a set of ideal sinusoids. In that case, if the amplitude spectrum as a function of frequency ω is $F(\omega)$ then the waveform is given by the Fourier integral,

$$f(t) = \frac{1}{2\pi} \int_{-\infty}^{\infty} F(\omega) \exp (j \omega t) \, d\omega \quad . \tag{1.4}$$

The individual sinusoids in the integrand of this equation are of infinite duration in time, whereas the waveform $f(t)$ that they represent is usually of finite duration. An immediate advantage of the Fourier integral of equation (1.4) is that it can be inverted to give

$$F(\omega) = \int_{-\infty}^{\infty} f(t) \exp (-j \omega t) \, dt \tag{1.5}$$

A similar advantage occurs in using the angular plane-wave spectrum to represent radiated fields, as the following outline analysis shows.

Suppose, instead of the cylindrical waves of Fig. 1.1, the field diffracted by a slit is represented as the superposition of a set of plane waves, of which a typical member is that shown in Fig. 1.2. The angle θ is now a variable describing the directions of the component plane waves constituting the angular spectrum.

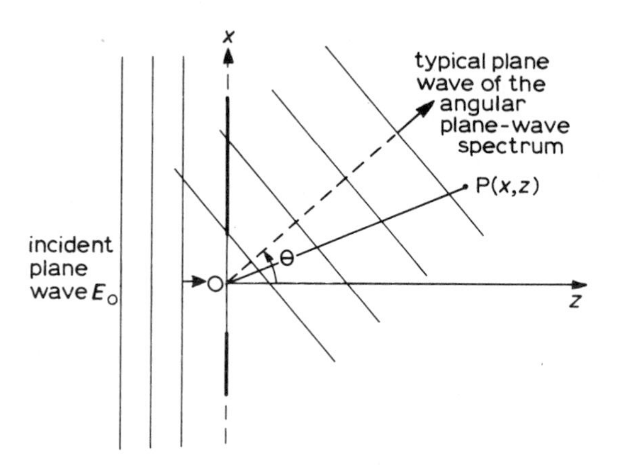

Fig. 1.2 – A plane wave diffracted by a slit. Plane-wave spectrum representation.

For technical reasons direction will be specified by sin θ rather than θ. Then if the set of plane waves is described by the spectrum function $F(\sin \theta)$, the contribution of this single plane wave of elemental amplitude $F(\sin \theta) d(\sin \theta)$ to the field at some point P is

$$dE_p(x,z) = F(\sin \theta) \, d(\sin \theta) \exp \{-jk \, (x \sin \theta + z \cos \theta)\} \tag{1.6}$$

The complete field at P is then obtained by integration as

$$E_p(x,z) = \int_{-\infty}^{\infty} F(\sin\theta) \exp\{-jk(x\sin\theta + z\cos\theta)\} d(\sin\theta) \tag{1.7}$$

(For present purposes it may be assumed that the range of integration is artificially extended beyond its natural limits of ± 1 for analytical convenience. However, it will be shown in Chapter 2 that waves travelling in directions such that $|\sin\theta| > 1$ are far from artifical). Now, defining the field over the aperture as

$$f(x) = E_p(x,0) \quad , \tag{1.8}$$

equations (1.7) and (1.8) can be combined to give

$$f(x) = \int_{-\infty}^{\infty} F(\sin\theta) \exp(-jkx\sin\theta)\; d(\sin\theta) \tag{1.9}$$

This is a Fourier integral that can be inverted to give the angular spectrum $F(\sin\theta)$ in terms of the aperture field $f(x)$ as

$$F(\sin\theta) = \frac{1}{\lambda}\int_{-\infty}^{\infty} f(x) \exp(jkx\sin\theta)\, dx \tag{1.10}$$

In the elementary diffraction example of section 1.1 the aperture field is assumed to be non-zero only across the slit, so that

$$f(x) = \begin{cases} E_0 \text{ for } |x| \leqslant a/2 \\ 0 \;\; \text{otherwise} \end{cases} \tag{1.11}$$

Substituting this into equation (1.10) yields the angular spectrum

$$F(\sin\theta) = \frac{E_0}{\lambda}\int_{-a/2}^{a/2} \exp(jkx\sin\theta)\, dx$$

$$= \frac{E_0 a}{\lambda} \frac{\sin\left(\dfrac{\pi a \sin\theta}{\lambda}\right)}{\left(\dfrac{\pi a \sin\theta}{\lambda}\right)} \tag{1.12}$$

which has the same angular dependence as the far field of equation (1.3) derived from Huygens' principle. Thus we can see in this particular example, and will later establish as generally true, that the angular spectrum gives the far field of the radiation. And furthermore, as equation (1.10) shows, the angular spectrum is simply the Fourier transform of the field across the radiating aperture. Once $F(\sin\theta)$ is known the radiated field $E_p(x,z)$ at any point to the right of the aperture plane is determined by the integral of equation (1.7).

The angular plane-wave spectrum representation was popular around the turn of the last century. Rayleigh (1896) used it to describe the fields reflected and transmitted at a corrugated dielectric boundary illuminated by a plane wave, and Debye (1909) used the representation to investigate the fields in the region of a focus. In 1902 Whittaker had proved that the field described by an angular spectrum of plane waves radiating in all directions was a solution of the wave equation. But this form of the angular spectrum leaves the source of radiation unspecified, and is therefore not unique. This limitation was overcome in Weyl's (1919) representation of the field radiating into a half-space bounded by a plane. The source field can be specified over this plane, and so the representation can be unique.

The application of the angular plane-wave spectrum to aperture antenna analysis and synthesis was pioneered by Booker and Clemmow (1950) and Woodward and Lawson (1948), essentially using the Weyl representation in two dimensions. Brown (1958) extended the representation to three dimensions, and introduced a reciprocity theorem for antennas which enabled the concept to be applied to receiving as well as transmitting antennas. This led to a transmitter/receiver coupling formula which applies to near-field coupling as well as to the far field. The angular plane-wave spectrum concept is now used extensively in planar-antenna synthesis (Rhodes, 1974) and in near-field antenna measurements (Paris and Joy, 1978).

In Chapters 2 and 3 of this book the concept of an angular spectrum of plane waves is established for two- and three-dimensional fields. In Chapter 4 the concept is applied to transmitting antennas, receiving antennas, and the coupling between them. Fresnel diffraction is examined in Chapter 5, and reflection from flat and curved conducting surfaces in Chapter 6. Particular examples of planar-aperture antennas are examined in detail in Chapter 7. Chapter 8 looks briefly at the problem of radiation from non-planar apertures. Appendices provide a summary of the properties of plane waves and of some theorems derived from Maxwell's equations. The remainder of this introductory chapter will be devoted to a physical description of the planar-aperture antennas to be examined in Chapter 7, and to giving a list of those performance characteristics and terms that are in common use in antenna practice.

1.3 SOME APERTURE ANTENNAS

The following examples of microwave antennas used in communications or radar can be described as planar-aperture antennas.

The Electromagntic Horn Antenna

At microwave frequencies a simple way of making a radiating antenna is to use an open-ended waveguide with a flared transition section (that is, a horn) added to achieve a reasonable match and increased directivity. (See Fig. 1.3).

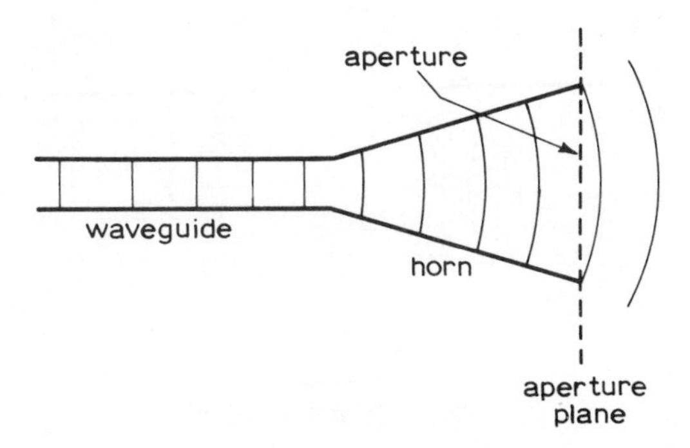

Fig. 1.3 – The Electromagnetic Horn Antenna. Feint lines indicate equiphase wavefronts of the radiating field.

An obvious starting point for the analysis of such a radiating device is to assume that the field over the mouth of the horn is an expanded form of the field distribution in the waveguide, but with a curved rather than plane wavefront. Thus the aperture of this antenna is the real aperture formed by the mouth of the horn, assumed to be part of the aperture plane. The field distribution over this plane, the aperture field distribution, could be taken approximately to be the expanded waveguide field over the actual aperture and zero over the remaining part of the aperture plane.

The electromagnetic horn is robust and fairly easy to manufacture. It is widely used as a standard in antenna gain measurements. (For a definition of antenna gain see section 1.4). But owing to the curvature of the emerging wavefront its gain is relatively low. The different methods that have been devised to correct the curved wavefront of simple primary sources, such as the electromagnetic horn, into the more desirable planar wavefront have led to a variety of antenna designs, of whcih the horn-lens combination is perhaps the most obvious.

The Horn-Lens Antenna

Here the curved wavefront of the field emerging from the mouth of the electromagnetic horn is corrected by the use of a converging lens. The resulting planar wavefront leads to an improvement in the directivity of the radiated field. Two

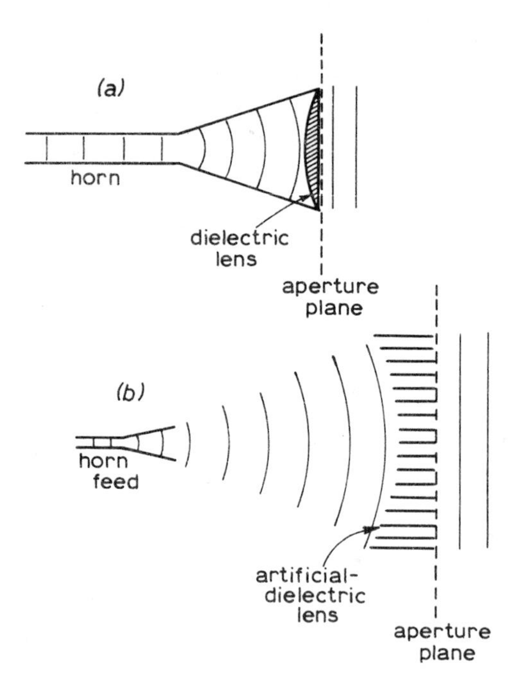

Fig. 1.4 – The Horn-Lens Antenna. The curved wavefront corrected with (a) a solid dielectric lens, and (b) an artificial-dielectric lens.

examples are shown in Fig. 1.4. The solid dielectric lens of Fig. 1.4(a) behaves in exactly the same way as a convex glass lens in optics. The artificial-dielectric lens shown schematically in Fig. 1.4(b) consists of a stack of metal-walled waveguides within which the phase velocity of the waves is greater than in air: hence the concave construction of the lens. In both cases the natural aperture is just beyond the output face of the lens.

The advantage of increased directivity and gain are somewhat offset in the horn-lens antenna by the reflections that inevitably occur at the front and back surfaces of the lens. The essential robustness of the electromagnetic horn is also diminished in the horn-lens combination.

The Horn-Paraboloid Antenna

An ingenious design which achieves the desired wavefront correction of the field emerging from a horn, without introducing partial reflections and without loss of robustness, is the horn-paraboloid assembly shown in Fig. 1.5. The metal cowl is welded on to the horn and has a paraboloidal (that is, part of a parabola of revolution) profile which transforms the spherical wavefront at the mouth of the horn into a planar wavefront which emerges from the side of the assembly. It is convenient in this case to suppose that the radiating aperture of this device lies just beyond the metal structure, as shown in the figure.

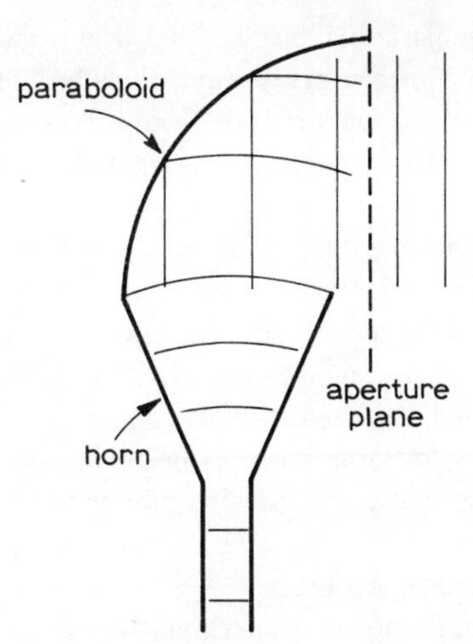

Fig. 1.5 – The Horn-Paraboloid Antenna.

However, there is another factor which affects antenna directivity. For maximum directivity the wavefront in the aperture should not only be planar, but as wide as possible. In the antenna designs so far mentioned emphasis has been on correcting the wavefront. We now turn to a series of antenna designs which both correct the wavefront and extend the lateral dimensions of the aperture.

The Paraboloid Reflector Antenna

This antenna consists of a primary feed, which is shown in Fig. 1.6 as an electromagnetic horn, positioned at the focus of a paraboloid reflector. The geo-

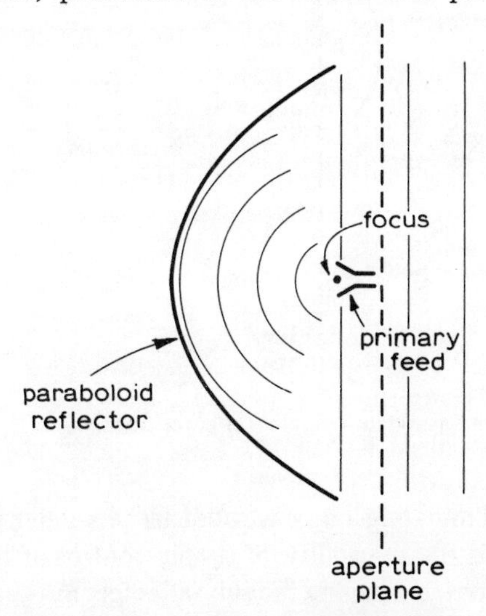

Fig. 1.6 – The Paraboloid Reflector Antenna.

metrical properties of the parabola ensure that the spherical wave incident from the feed is transformed into a plane wave on reflection. The lateral extent of the emerging aperture field, which it is convenient to place beyond the primary feed and its support structure, can be as wide as the lateral extent of the paraboloid.

The simplicity of design (basically that of the searchlight and Newtonian telescope) and its robustness of construction make the focus-fed paraboloid reflector a very popular choice as a high-gain antenna in communications and radar. However, the physical presence between the reflector and the supposed aperture of the primary feed, its feeder waveguide and its associated support structure, gives rise to some deterioration in performance, and to some rather unsatisfactory modifications to the analysis, in comparison with the ideal.

The Cassegrain Double Reflector Antenna

The addition of a second reflector of hyperboloidal shape, labelled as the sub-reflector in Fig. 1.7, leads to an even more robust and compact design, known as the Cassegrain. (The Cassegrain system also originated as a design for an optical telescope). The primary feed is inserted through the centre of the main

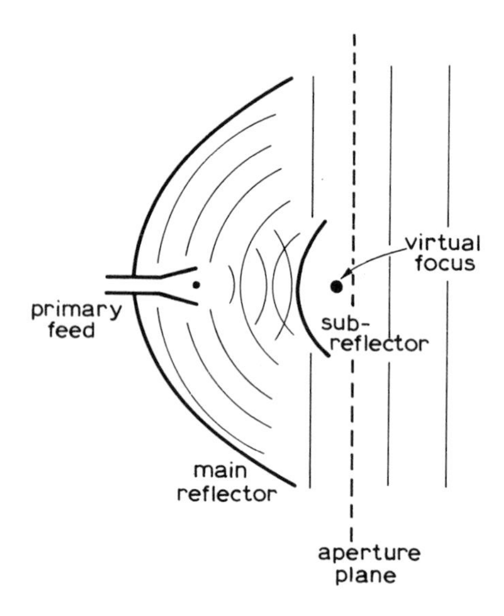

Fig. 1.7 – The Cassegrain Double Reflector Antenna.

paraboloidal reflector, thus eliminating long waveguide feeders and their associated noise problems. There is also the possibility of simple control of the antenna's radiation pattern by appropriately shaping the sub-reflector. But the problem of aperture blocking, this time by the sub-reflector and its supports, still remains.

The Offset-Fed Paraboloid Reflector Antenna

The problem of blocking of the desired aperture field by the feed can be largely overcome by placing the primary feed in an offset position, as indicated in Fig. 1.8, and appropriately restricting the extent of the main reflector. The price to be paid for the resulting improvement is that such an antenna is awkward to manufacture (at least for mass production), and the skew geometry can give rise to increased cross-polarized radiation, which may be undesirable.

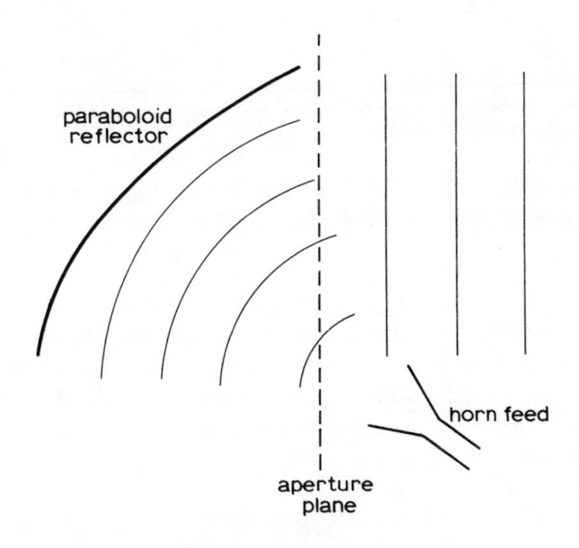

Fig. 1.8 – The Offset-Fed Paraboloid Reflector Antenna.

Nonetheless, our main concern here is not with the pros and cons of different antennas, but with the common feature that makes them eligible to be considered as aperture antennas; which is that an analysis of their performance arises naturally (but not necessarily exclusively) from consideration of a field distribution over some surface in the vicinity of the antenna structure. This is clearly so in the present instance, as suggested in Fig. 1.8.

1.4 BASIC ANTENNA CONCEPTS

The object of antenna analysis is an accurate description of the radiating and receiving characteristics of an antenna. Over the years certain concepts and definitions have developed and become established as part of the common language among antenna engineers. We will review these concepts and definitions here. They will be developed in detail and discussed at greater length in later chapters.

The Radiated Far Field

At a large distance r from any antenna its electric field can be expressed in the form

$$\mathbf{E}(r,\theta,\phi) = \frac{\exp(-jkr)}{kr} \ \mathbf{e}(\theta,\phi) \tag{1.13a}$$

such that

$$\mathbf{u}_r \,.\, \mathbf{E}(r,\theta,\phi) = 0 \quad , \tag{1.13b}$$

and the associated magnetic field is

$$\mathbf{H}(r,\theta,\phi) = \frac{1}{Z} \ \mathbf{u}_r \times \mathbf{E}(r,\theta,\phi) \tag{1.14}$$

The observation point P (See Fig. 1.9) is assumed to lie on a sphere of radius r, centred on a convenient point O in the vicinity of the antenna. The spherical polar coordinates of the point P are (r,θ,ϕ) where θ, the polar angle, and ϕ, the azimuth angle, together define the direction of P from O. The direction (θ,ϕ) is also that of the unit vector in the radial direction, $\mathbf{u}_r$.

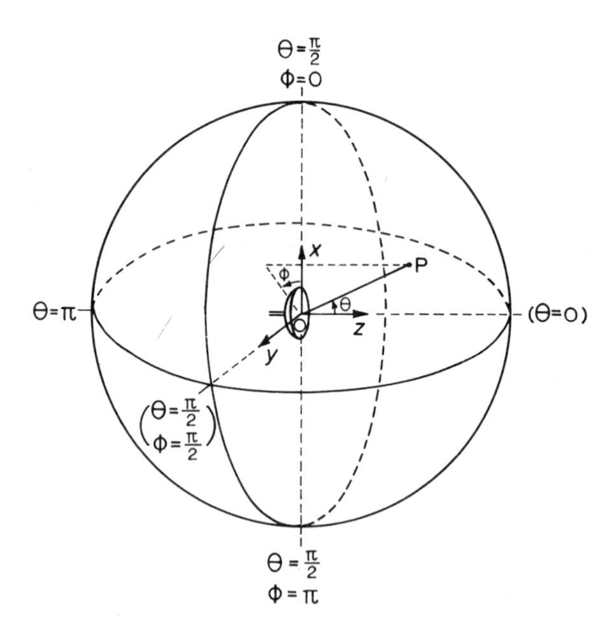

Fig. 1.9 – Geometry for the radiated field.

The following features of the antenna far field should be noted:

(a) The distance r of the observation point from the antenna should be at least the Rayleigh distance, which will be shown in section 5.5.3 to be $2a^2/\lambda$, where a is the maximum dimension of the radiating aperture.

(b) The time variation of the fields is understood to be sinusoidal, having the complex form $\exp(j\omega t)$. This means that the actual space-time behaviour for a phasor-vector such as $\mathbf{E}(r,\theta,\phi)$ is obtained from

$$\mathbf{E}(r,\theta,\phi,t) = \text{Re}\ [\mathbf{E}(r,\theta,\phi)\, e^{j\omega t}] \tag{1.15}$$

where Re denotes the real part. The convention we are using is that the phasor-vector $\mathbf{E}$ has a complex magnitude E and an absolute magnitude $|E|$, which is also the peak value of the sinusoidal time variation.

(c) The form of the far electric field $\mathbf{E}(r,\theta,\phi)$ of equation (1.13a) is the product of a uniform spherical scalar wave and the vector pattern function $\mathbf{e}(\theta,\phi)$.

(d) The dimensions of $\mathbf{e}(\theta,\phi)$ are the same as those of $\mathbf{E}(r,\theta,\phi)$, namely volts per metre, as a consequence of arbitrarily multiplying r in the denominator of equation (1.13a) by the plane-wave phase constant $k = \omega\sqrt{\mu\epsilon}$.

(e) In a particular direction (θ,ϕ) the amplitude of the field falls off as r^{-1}, and its phase is retarded linearly as kr, with increasing radial distance r.

(f) At a constant radial distance the dependence on direction of the amplitude, phase and polarization (that is, vector direction) of the electric field is given by $\mathbf{e}(\theta,\phi)$, which is the vector pattern function for a particular antenna.

(g) The electric field $\mathbf{E}(r,\theta,\phi)$ is polarized such that it is always tangential to the sphere of radius r, as specified by equation (1.13b) This is illustrated in Fig. 1.10.

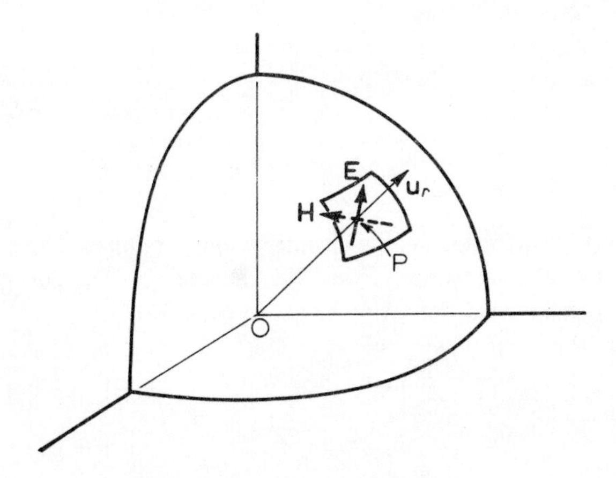

Fig. 1.10 – Polarization of the far field. E,H, and $\mathbf{u}_r$ form a right-handed set.

(h) The magnetic field $\mathbf{H}(r,\theta,\phi)$ of equation (1.14) is also tangential to the sphere of radius r. It is also orthogonal to $\mathbf{E}(r,\theta,\phi)$. Thus $\mathbf{E}$,$\mathbf{H}$ and $\mathbf{u}_r$ are mutually orthogonal everywhere in the far field. They form a right-handed set in the stated order (that is, if $\mathbf{E}$ is rotated on to $\mathbf{H}$, a right-hand screw thread would advance in the direction of $\mathbf{u}_r$) as specified by equations (1.13) and (1.14) and illustrated in Fig. 1.10.

(i) In equation (1.14), which relates the electric and magnetic fields, Z is the plane-wave impedance of the propagating medium. In terms of the permeability μ and permittivity ϵ of the medium,

$$Z = \sqrt{\frac{\mu}{\epsilon}} \tag{1.16}$$

(j) In the neighbourhood of any point **P** in the field the electric and magnetic fields have the *local* character of a plane wave travelling in the radial direction.

(k) The power flux density, given by half the real part of the complex Poynting vector

$$\mathbf{S} = \tfrac{1}{2}\,\mathrm{Re}\,\mathbf{E} \times \mathbf{H}^* \quad , \tag{1.17}$$

is directed radially outward (that is, $\mathbf{S} = \mathbf{u}_r S_r$) for the values of $\mathbf{E}$ and $\mathbf{H}$ given above for the far field. The asterisk denotes complex conjugate. The radial component of the power flux density is therefore

$$S_r(r,\theta,\phi) = \frac{\mathbf{e}(\theta,\phi).\,\mathbf{e}^*(\theta,\phi)}{2(kr)^2 Z} = \frac{|e(\theta,\phi)|^2}{2(kr)^2 Z} \tag{1.18}$$

Directivity, Gain and Efficiency

The directivity $D(\theta,\phi)$ of a radiating antenna is defined as the ratio:

$$D(\theta,\phi) = \frac{\text{Power flux density (p.f.d.) from antenna in direction } (\theta,\phi)}{\text{P.f.d. when same power is radiated uniformly in all directions}} \; . \tag{1.19}$$

The power flux densities in this definition must be measured at the same radial distance, or equivalently the power flux density must be defined per unit solid angle. In either case the definition refers to the far field.

If P_t is the total power radiated by the antenna, the power flux density at a distance r when this is radiated uniformly in all directions (that is, radiated isotropically) is

$$S_r^{\mathrm{iso}}(r,\theta,\phi) = \frac{P_t}{4\pi r^2} \quad , \tag{1.20}$$

and the antenna directivity is

$$D(\theta,\phi) = \frac{S_r(r,\theta,\phi)}{S_r^{iso}(r,\theta,\phi)} = \frac{\lambda^2 |e(\theta,\phi)|^2}{2\pi Z P_t} \tag{1.21}$$

for the far fields defined earlier.

Antenna gain $G(\theta,\phi)$ has a definition similar to that of directivity, the only difference being that the denominator of equation (1.19) is based on the input power P_{in} delivered to the antenna. Thus

$$G(\theta,\phi) = \eta\, D(\theta,\phi) \quad ; \quad 0 \leqslant \eta \leqslant 1 \quad , \tag{1.22}$$

where η, known as the antenna efficiency, is the ratio of total power radiated to the power input to the antenna, namely,

$$\eta = P_t / P_{in} \tag{1.23}$$

In terms of the far-field pattern function, then,

$$G(\theta,\phi) = \frac{\lambda^2 |e(\theta,\phi)|^2}{2\pi Z P_{in}} \tag{1.24}$$

The maximum value of the gain function $G(\theta,\phi)$ is also often referred to simply as the gain of the antenna.

The polarization of the far field is described by the complex vector $\mathbf{e}(\theta,\phi)$, which is constrained to lie in a plane perpendicular to the radial direction. The vector $\mathbf{e}$ can therefore be resolved into two components with basis vectors $\mathbf{e}_1$ and $\mathbf{e}_2$, so that

$$\mathbf{e} = a_1 \mathbf{e}_1 + a_2 \mathbf{e}_2 \tag{1.25}$$

in which a_1 and a_2 are complex. The basis vectors can be so chosen that they are normalised with

$$\mathbf{e}_1 . \mathbf{e}_1{}^* = \mathbf{e}_2 . \mathbf{e}_2{}^* = 1 \tag{1.26}$$

and orthogonal in the sense that

$$\mathbf{e}_1 . \mathbf{e}_2{}^* = \mathbf{e}_1{}^* . \mathbf{e}_2 = 0 \quad . \tag{1.27}$$

This resolution of the field could be into two orthogonal linearly polarized plane waves, or into a combination of right-hand and left-hand circularly polarized plane waves, whichever is more appropriate. Then

$$\mathbf{e} . \mathbf{e}^* = |a_1|^2 + |a_2|^2 = |e|^2 \tag{1.28}$$

and it is clear from equation (1.21) that the directivity can be resolved into two parts, namely,

$$D(\theta,\phi) = D_1(\theta,\phi) + D_2(\theta,\phi) \tag{1.29}$$

based on the chosen resolution of the field. Equation (1.24) shows that the gain $G(\theta,\phi)$ can be similarly resolved.

Radiation Patterns, Beamwidth and Sidelobes

It is usually required of an antenna that it be directive. The simplest measure of its effectiveness as such is the gain, that is, the maximum value of the gain function $G(\theta,\phi)$. The gain is thus the ratio, invariably stated in decibels (dB), of the maximum power flux density produced by the antenna to its value if the power delivered to the antenna had been radiated isotropically.

More information about its directive properties can be obtained from the antenna's radiation patterns. These are plots of radiated field strength or, more usually, power flux density (directivity or gain) as a function of angle. Since direction is specified by two angles whereas it is only possible to plot against one, antenna radiation patterns consist of a set of sections in the $\theta-\phi$, or some equivalent, plane. For example, suppose that an antenna has its maximum power radiation in the direction $\theta = 0$. Then a plot of the antenna gain, in decibels relative to the maximum value, as a function of θ with ϕ held constant might be that sketched in Fig. 1.11. Two methods of plotting the same information are shown, one in polar the other in rectangular form.

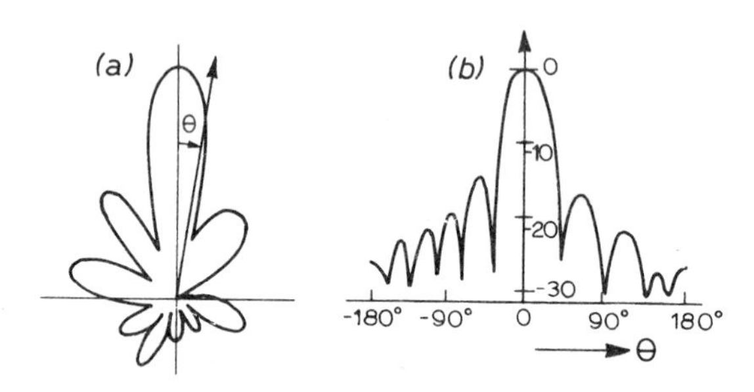

Fig. 1.11 – Polar and rectangular plots of relative gain in dB versus θ for constant ϕ.

Antenna radiation patterns often have a lobe structure similar to that shown. The lobe containing the direction of maximum radiated power is the main lobe; all other lobes are referred to as sidelobes. The two sidelobes adjacent to the main lobe are the first sidelobes, and the lobe diametrically opposite the main lobe, if it exists, is called the back lobe. The lobes are separated by nulls, so called because the power radiated in these directions can in theory be zero.

The angular width of the main lobe of the antenna, known as its beam width, is a useful measure of the capability of the antenna of resolving the angular position of a distant point source. Two common ways of defining antenna beamwidth are shown in Fig. 1.12. One is the angle between the two points on either side of the main lobe at which the radiated power has fallen to half its maximum value, that is, the -3 dB points, and is known as the 3-dB beamwidth. The other definition is the angle between the first nulls of the pattern, which is usually about twice the 3-dB beamwidth. This fact makes the additional point that if two distant sources are separated in angle by the 3-dB beamwidth and one of them lies in the centre of the main beam, then the other will lie in the region of the first null.

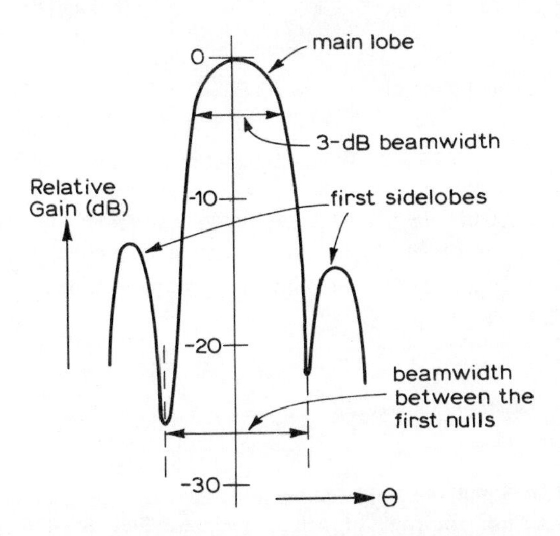

Fig. 1.12 – Details of the main lobe and first sidelobes of the antenna pattern in Fig. 1.11(b), showing two ways of defining antenna beamwidth.

Receiving Antennas

The directivity of a receiving antenna is defined in terms of its response to an incident plane wave. It takes the form of an effective receiving area:

$$A(\theta,\phi) = \frac{\text{Power delivered to receiver}}{\text{Power density in plane wave incident from direction } (\theta,\phi)} \tag{1.30}$$

It is assumed that the polarization of the incident plane wave is adjusted to deliver maximum power to the receiver: a condition known as **polarization match.**

Consider the plane wave to be $\mathbf{E}'$, incident from the direction (θ',ϕ'), as shown in Fig. 1.13. If the antenna has a vector far-field radiation pattern $\mathbf{e}(\theta,\phi)$, the antenna reciprocity theorem of section 4.2.1 states that the received signal will be proportional to $\mathbf{E}'.\mathbf{e}(\theta',\phi')$. The vectors $\mathbf{E}'$ and $\mathbf{e}(\theta',\phi')$ lie in the same plane: the condition of polarization match is therefore when they are coincident in that plane. When this occurs the power delivered to the receiver will be proportional to $|E'|^2.|e(\theta',\phi')|^2$, which means that it is proportional to the product

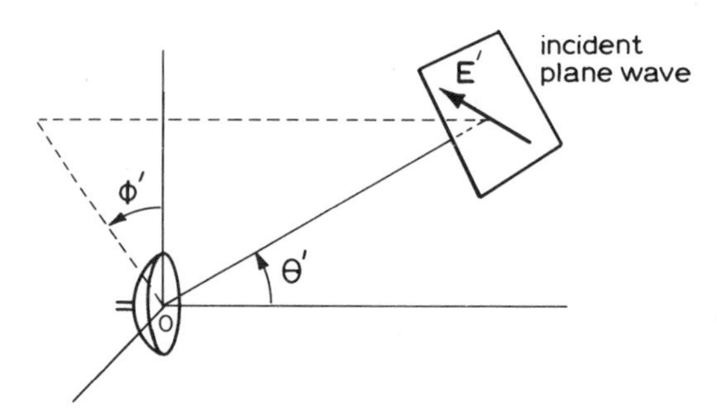

Fig. 1.13 – Plane wave incident on a receiving antenna.

of the power density in the incident plane wave and the gain of the antenna given by equation (1.24). Thus the effective receiving area of an antenna is proportional to its transmitting gain. The precise relationship will be shown in section 4.2.2 to be

$$A(\theta,\phi) = \frac{\lambda^2}{4\pi} G(\theta,\phi) \tag{1.31}$$

Transmitter to Receiver Coupling

An immediate use for the definitions we have just given of antenna gain and receiving area is the far-field coupling equation, known as the **Friis transmission formula** which we will now derive. (This is a particular form of a more general result that will be obtained later in section 4.3).

Suppose that the transmitting and receiving antennas are disposed as shown in Fig. 1.14. The distance r between them is presumed to satisfy the far-field criterion that $r \geqslant 2a^2/\lambda$, where a is now the largest dimension of both antenna

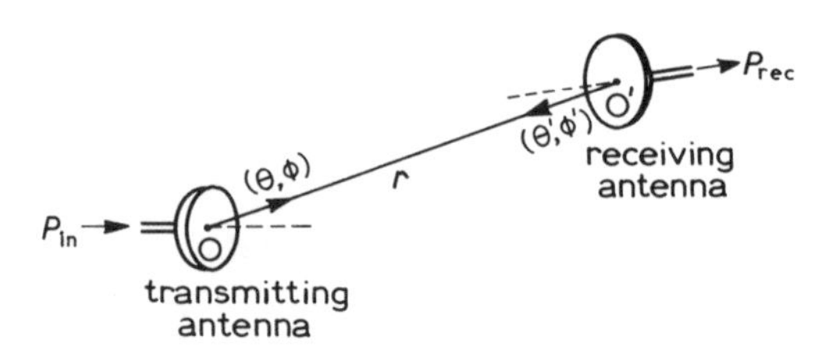

Fig. 1.14 – Direct coupling between a transmitting and receiving antenna.

apertures. The receiver is in the direction (θ,ϕ) from the transmitter, and the transmitter is in the direction (θ',ϕ') from the receiver ;and if the transmitting antenna gain is $G_T(\theta,\phi)$, when P_{in} is delivered to the transmitter, the power flux density incident on the receiving antenna in the form of a locally plane wave will be

$$S_r = P_{in}\frac{G_T(\theta,\phi)}{4\pi r^2} \tag{1.32}$$

If the effective receiving area of the receiving antenna is $A_R(\theta',\phi')$, the power delivered to the receiver will be

$$P_{rec} = P_{in}\frac{G_T(\theta,\phi)\,A_R(\theta',\phi')}{4\pi r^2}\,, \tag{1.33}$$

it being assumed that the receiving antenna is polarization matched to the incident field. An alternative form for the received power, using equation (1.31) is

$$P_{rec} = P_{in}\frac{\lambda^2 G_T(\theta,\phi)\,G_R(\theta',\phi')}{(4\pi)^2 r^2} \tag{1.34}$$

Another important form of coupling between a transmitter and a reciever occurs indirectly via a scattering object, as in radar. The scatterer is represented by its **scattering cross-section,** which has the following rather tortuous but basically simple definition. The scattering cross-section σ is that area which, when placed perpendicular to the plane-wave field incident on the scatterer, would intercept that amount of power which, when radiated uniformly in all directions, would give the observed power flux density in a particular direction. It is therefore a function of the direction and polarization of the incident field, and of the direction of observation.

If, with the geometry shown in Fig. 1.15, power P_{in} is delivered to the

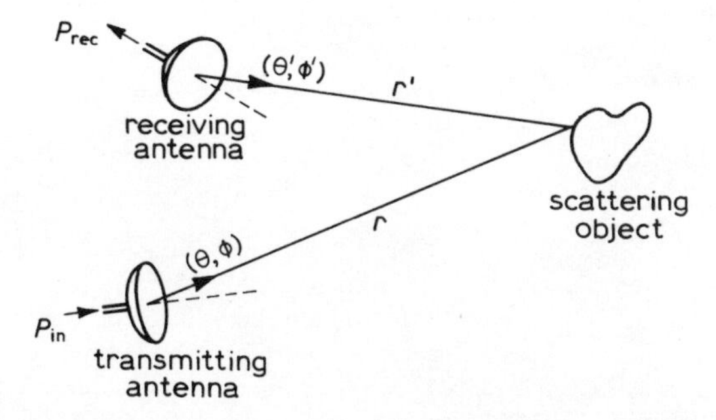

Fig. 1.15 – Geometry for the indirect coupling between a transmitting and receiving antenna via a scattering object.

transmitting antenna and the scatterer is in its far field, the power intercepted will be $P_{in}\, G_T(\theta,\phi)\, \sigma/4\pi r^2$, and the power delivered to the receiver will therefore be

$$P_{rec} = P_{in} \frac{\sigma G_T(\theta,\phi) A_R(\theta',\phi')}{(4\pi)^2 r^2 r'^2} \quad , \tag{1.35}$$

again assuming that the receiving antenna is polarization matched to the field incident upon it. In radar systems the same antenna is often used for transmitting and receiving, in which case the received power is

$$P_{rec} = P_{in} \frac{\sigma \lambda^2 G^2(\theta,\phi)}{(4\pi)^3 r^4} \quad , \tag{1.36}$$

where we have set $G_T\,(\theta,\phi) = G_R\,(\theta,\phi) = G(\theta,\phi)$, $r' = r$ and have made use of equation (1.31).

CHAPTER 2

Plane-wave representation of two-dimensional fields

Plane waves are exact solutions of Maxwell's fundamental field equations. Maxwell's equations are linear, provided the medium itself is linear (that is, the electrical properties of the medium such as its permittivity and permeability are independent of the electric and magnetic fields). Hence the principle of linear superposition applies, and the superposition of individual plane waves travelling in different directions in the medium constitutes an exact solution.

Our ultimate goal is to study the diffraction of electromagnetic waves in realistic, three-dimensional situations. But it is much easier to introduce the underlying concepts and techniques in two dimensions. Hence this chapter is devoted to building up a picture of two-dimensional fields in terms of individual plane waves travelling in different directions, and to examining the implications of this representation. In the following chapter it will then be possible to extend the representation to three dimensions, concentrating on the inevitably increased complexity of the mathematics while taking the underlying concepts for granted.

2.1 PLANE WAVES

We begin by listing the properties of homogeneous plane electromagnetic waves. These will be familiar to most readers. But if they are not, a derivation based on transmission-line theory can be found in Appendix A.

A plane electromagnetic wave, of single-frequency time dependence $\exp(j\omega t)$, travelling in a uniform, isotropic and lossless open region, has the following properties:

(a) It is characterized by a single direction, such as that denoted by the unit vector $\mathbf{u}$ in Fig. 2.1. This direction is the direction of propagation of the plane wave.

(b) Over planes at right angles to $\mathbf{u}$ the phasor electric and magnetic fields, E and H, are constant in both amplitude and phase.

(c) The complex amplitudes E and H are related to each other by the proportionality

$$E = ZH \tag{2.1}$$

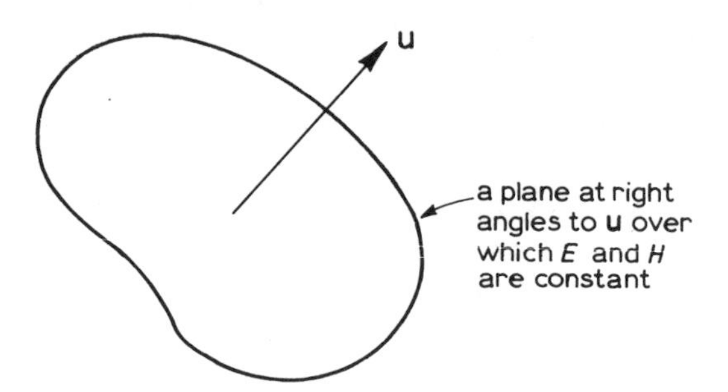

Fig. 2.1 – A plane wave travelling in the direction of **u**.

in which the constant Z is the characteristic (or plane-wave) wave impedance of the medium. In the SI system, the units of E are volts per metre and those of H are amperes per metre, so that the units of Z are ohms. In a uniform lossless medium of permeability μ and permittivity ϵ

$$Z = (\mu/\epsilon)^{1/2} \tag{2.2}$$

and is therefore real. This means that the electric and magnetic fields will be in phase. If the plane wave is travelling in free space, $\mu = 4\pi \times 10^{-7}$ henries per metre and $\epsilon = 8.854 \times 10^{-12}$ farads per metre, and its characteristic impedance is 376.7 ohms.

(d) The directions of the vector electric field **E**, the vector magnetic field **H**, and the direction of propagation **u** are mutually at right angles, as indicated in Fig. 2.2. The vectors (**E**,**H**,**u**) form a right-handed set, which means that they bear the same relationship to each other as, for example, the directions ($\mathbf{u}_x$,$\mathbf{u}_y$,$\mathbf{u}_z$) of the axes of the usual Cartesian coordinate system.

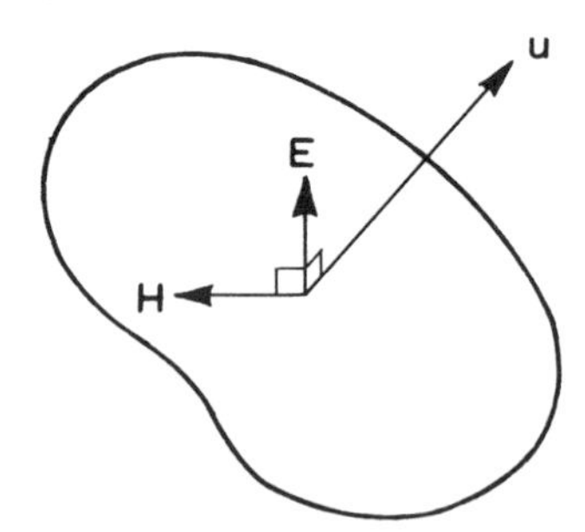

Fig. 2.2 – Vector fields **E** and **H** form a right-handed set with **u**.

Exercise

Suppose that the above plane wave is travelling in free space in a direction from the origin of a Cartesian coordinate system towards the point (1,1,1), and that we are told that the x-component of the electric field has a peak value of 10^{-3}

Vm^{-1}, and that the y-component of the electric field is zero. Calculate the remaining field components.

Exercise

Show that for the plane wave described geometrically in Fig. 2.2 the vector magnetic field is given by the vector

$$\mathbf{H} = Z^{-1}\mathbf{u} \times \mathbf{E} \tag{2.3}$$

(e) In a lossless medium the magnitudes of the fields are the same everywhere, but the phase of both the electric and magnetic fields is retarded in the direction of propagation of the wave. Thus, if the electric field over some reference plane

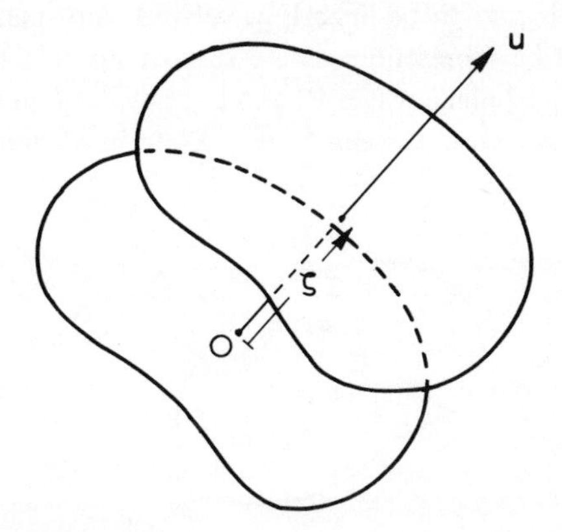

Fig. 2.3 – Planes separated by a distance ζ in the direction of propagation.

containing the point O in Fig. 2.3 is E_o, then the electric field E over a plane a distance ζ away from the reference plane in the direction of propagation will be

$$E = E_o \exp(-jk\zeta) \tag{2.4}$$

where the time factor exp $(j\omega t)$, as is usual for phasor representation of fields, has been suppressed. The phase constant k is the amount by which the phase of the plane wave is retarded in unit distance, and is given by

$$k = \omega(\mu\epsilon)^{1/2} \tag{2.5}$$

for a lossless medium of permeability μ and permittivity ϵ. The wavelength λ of a propagating wave is the distance over which the phase changes by 2π. Hence the phase constant can also be written as

$$k = 2\pi/\lambda \tag{2.6}$$

Exercise

Find the characteristic impedance and speed of travel of a plane wave propagating in a uniform lossless dielectric of relative permeability 1 and relative permittivity 2.56.

If now the same dielectric has a small amount of conductivity such that its loss tangent (for a definition see Appendix A.6) is 0.02 instead of zero, how will this change the characteristic impedance and speed of travel of the plane wave? What is the loss in dB/km?

(f) In antenna parlance the term polarization refers to the direction of the vector electric field, which for a plane wave is necessarily confined to planes orthogonal to **u**. If **E** points in a single direction throughout all time and space, then the plane wave is said to be **linearly polarized.** Any plane wave travelling in the direction **u** can be represented as the sum of suitable scalar multiples of any two linearly polarized plane waves $\mathbf{E}_1$ and $\mathbf{E}_2$, provided they are not collinear. The most convenient choice is to take $\mathbf{E}_1$ and $\mathbf{E}_2$ to be at right angles, as shown in Fig. 2.4.

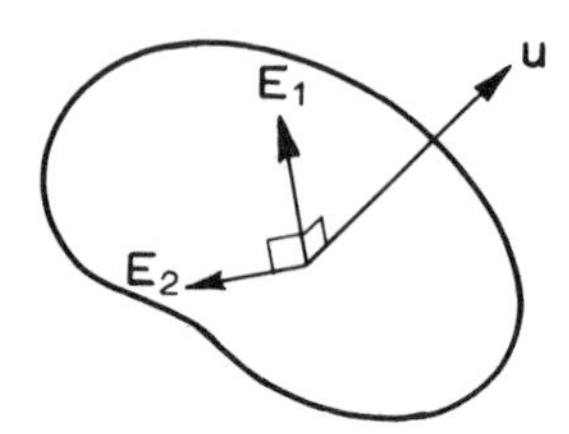

Fig. 2.4 – Two orthogonal linearly polarized plane waves $\mathbf{E}_1$ and $\mathbf{E}_2$.

A **circularly polarized** plane wave is one which can be represented as the sum of two orthogonal linearly polarized plane waves of equal amplitude but out of phase by $\pi/2$ radians. Let $\mathbf{E}_1 = E_1\,\mathbf{u}_1$ and $\mathbf{E}_2 = E_2\,\mathbf{u}_2$, where $\mathbf{u}_1$ and $\mathbf{u}_2$ are orthogonal unit vectors, and put $E_1 = E_o$ and $E_2 = \mp j\,E_o$. Then the plane wave with electric field

$$\mathbf{E} = E_o\,\mathbf{u}_1 \mp j\,E_o\,\mathbf{u}_2 \tag{2.7}$$

is a circularly polarized plane wave, the negative sign corresponding to clockwise rotation of the electric vector as the wave propagates (also known as right-handed circular polarization), and the positive sign corresponding to anticlockwise rotation (or left-handed circular polarization).

If the two orthogonal linearly polarized plane waves are combined in phase, then the resultant wave is again linearly polarized. The combination of two such orthogonal linearly polarized plane waves is akin to the formation of Lissajou figures on an oscilloscope screen. When their amplitudes and relative phases are arbitrary the tip of the electric-field vector traces out an ellipse, and the plane wave is said to be **elliptically polarized.** (see Appendix A.5 for a fuller treatment).

Exercise
Show that a linearly polarized plane wave can be represented as the sum of two circularly polarized plane waves travelling in the same direction, of the same amplitude but of opposite sense. Hence show that an arbitrarily polarized plane wave can be represented as a suitable weighted sum of two circularly polarized waves of opposite sense.

Exercise
The two linearly polarized electric field vectors $\mathbf{E}_1$ and $\mathbf{E}_2$ depicted in Fig. 2.4 satisfy the general orthogonality condition for electromagnetic fields, which is that the scalar product

$$\mathbf{E}_1.\mathbf{E}_2{}^* = 0 \tag{2.8}$$

in which the asterisk denotes the complex conjugate. Form the two circularly polarized plane waves

$$\mathbf{E}_r = E_1\,\mathbf{u}_1 - j\,E_1\,\mathbf{u}_2 \text{ and } \mathbf{E}_\ell = E_2\,\mathbf{u}_1 + j\,E_2\,\mathbf{u}_2 \tag{2.9}$$

in which $\mathbf{u}_1$ and $\mathbf{u}_2$ are spatially orthogonal, and show that $\mathbf{E}_r$ and $\mathbf{E}_\ell$ are themselves orthogonal.

(g) The power flow in the plane wave is given, as for any electromagnetic wave, by the vector

$$\mathbf{S} = \tfrac{1}{2}\,\mathrm{Re}\,\mathbf{E} \times \mathbf{H}^* \tag{2.10}$$

which is the Poynting vector averaged over one cycle of the oscillation of frequency $f = 2\pi/\omega$. The asterisk denotes complex conjugate and Re the real part. It is clear from Fig. 2.2 that the Poynting vector $\mathbf{S}$ is in the same direction as $\mathbf{u}$, the direction of propagation of the plane wave. The units of $\mathbf{S}$ are watts per square metre. For the homogeneous plane wave of Fig. 2.2 in a lossless medium, the electric and magnetic fields are not only spatially orthogonal but are also precisely in phase, so the vector power flow is

$$\mathbf{S} = (2Z)^{-1}\,|E|^2\,\mathbf{u} = (Z/2)|H|^2\mathbf{u} \tag{2.11}$$

Exercise
Rederive the last formula for the power flow in a plane wave by substituting the vector relationship between $\mathbf{E}$ and $\mathbf{H}$, namely

$$\mathbf{H} = Z^{-1}\,\mathbf{u} \times \mathbf{E} \text{ or } \mathbf{E} = Z\,\mathbf{H} \times \mathbf{u} \tag{2.12}$$

into the formula for the average Poynting vector. Use the vector triple product relations for any three vectors **a**, **b** and **c**, that

$$\mathbf{a} \times (\mathbf{b} \times \mathbf{c}) = (\mathbf{a}.\mathbf{c})\mathbf{b} - (\mathbf{a}.\mathbf{b})\mathbf{c}$$

and

$$(\mathbf{a} \times \mathbf{b}) \times \mathbf{c} = (\mathbf{c}.\mathbf{a})\mathbf{b} - (\mathbf{c}.\mathbf{b})\mathbf{a} \tag{2.13}$$

Exercise

Show that the power flow in a plane wave represented as the sum of two orthogonally polarized plane waves is just the sum of the power flows in the two waves.

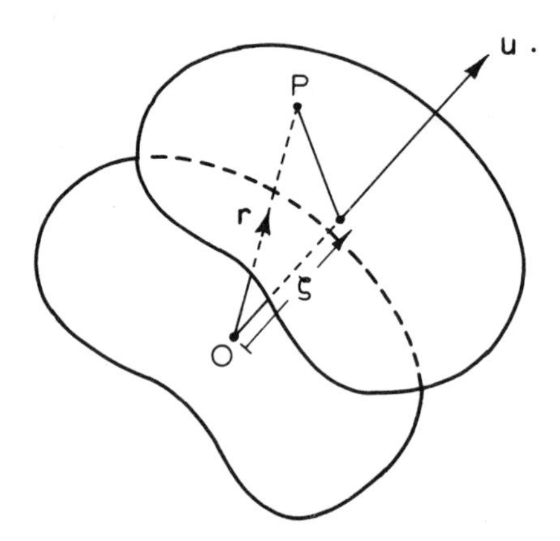

Fig. 2.5 – Geometry to determine the field at P when a plane wave has electric field $\mathbf{E}_o$ over the plane containing the point O.

(h) Collecting the properties of plane waves given in sections (a) to (g), if a plane wave is propagating in the direction **u** in a lossless, uniform, isotropic and unbounded medium, and if its vector electric field is $\mathbf{E}_o$ over the plane containing the point O (see Fig. 2.5), then the vector electric field at the point P, whose vector position with respect to the point O is **r**, will be

$$\mathbf{E}(\mathbf{r}) = \mathbf{E}_o \exp(-jk\,\mathbf{u}.\mathbf{r}) \tag{2.14}$$

since the distance between the planes containing O and P is $\zeta = \mathbf{u}.\mathbf{r}$. The electric field, whatever the polarization, must be orthogonal to **u**, which is expressed by

$$\mathbf{u}.\mathbf{E}(\mathbf{r}) = 0 \tag{2.15}$$

The vector magnetic field must be orthogonal to both **u** and **E(r)**, and its phasor amplitude is related to that of **E(r)** by the characteristic impedance Z of the medium, so that

$$\begin{aligned}\mathbf{H}(\mathbf{r}) &= Z^{-1}\,\mathbf{u} \times \mathbf{E}(\mathbf{r}) \\ &= Z^{-1}\,\mathbf{u} \times \mathbf{E}_o \exp(-jk\,\mathbf{u}.\mathbf{r})\end{aligned} \tag{2.16}$$

The corresponding expression for the power flow in the plane wave is

$$\begin{aligned}\mathbf{S}(\mathbf{r}) &= \tfrac{1}{2}\,\mathrm{Re}\,\mathbf{E}(\mathbf{r}) \times \mathbf{H}^*(\mathbf{r}) = \tfrac{1}{2}\,\mathrm{Re}\,\mathbf{E}_o \times Z^{-1}(\mathbf{u} \times \mathbf{E}_o^*)\\ &= (2Z)^{-1}\,(\mathbf{E}_o.\mathbf{E}_o^*)\,\mathbf{u}\\ &= (2Z)^{-1}\,|E_o|^2\,\mathbf{u}\end{aligned} \tag{2.17}$$

Some authors prefer to combine the phase constant k and direction $\mathbf{u}$ into the single quantity

$$\mathbf{k} = k\mathbf{u}, \tag{2.18}$$

sometimes known as the **vector wavenumber** of the plane wave. It has the virtue of giving information about the frequency (since $k = \omega(\mu\epsilon)^{1/2}$) and direction of the plane wave in one symbol. The electric field at point P is then

$$\mathbf{E}(\mathbf{r}) = \mathbf{E}_o \exp(-j\mathbf{k}.\mathbf{r}) \tag{2.19}$$

and similarly for the magnetic field. This slight simplification in the argument of the exponential term is offset by increased complexity elsewhere: in particular, wherever $\mathbf{u}$ occurs it must be replaced by $\mathbf{k}/k$. We have preferred to retain the explicit dependences of field expressions on the direction $\mathbf{u}$, or something equivalent to it, partly for this reason. But the main reason for our choice is to emphasize the fact that the angular plane-wave spectrum, which we are now in a position to examine in detail, is a function of direction.

2.2 ANGULAR SPECTRUM FOR TWO-DIMENSIONAL FIELDS

Referring to the Cartesian coordinate system of Fig. 2.6, it will be supposed that the fields are uniform with y, and hence depend on the coordinates (x,z)

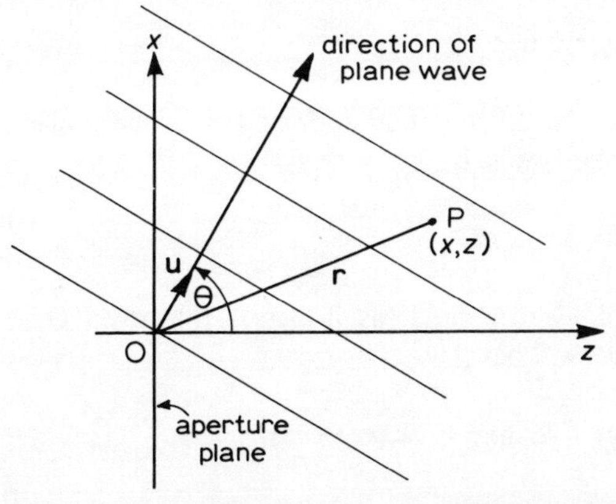

Fig. 2.6 – Two-dimensional geometry for the fields which are uniform with y, which is at right angles to the plane of the figure.

only. The x-y plane will be taken to be the aperture plane, and our interest will be in the fields diffracted into the supposed uniform, isotropic, source-free and lossless medium to the right of the aperture plane, the half-space $z \geqslant 0$.

To construct the fields in this half-space from a set of plane waves travelling in different directions, it is clear that they must all be travelling parallel to the x-z plane in order that the fields are independent of y. A typical member of the set of plane waves is shown in Fig. 2.6, its direction $\mathbf{u}$ making an angle θ to the z axis. Since the sources are to the left of the x-y aperture plane, the plane waves must all travel into the half-space $z \geqslant 0$, which restricts θ to the range

$$-\pi/2 \leqslant \theta \leqslant \pi/2 \ , \tag{2.20}$$

assuming for now that θ is real.

The vector position of the field point P with respect to the origin O is

$$\mathbf{r} = \mathbf{u}_x x + \mathbf{u}_z z \tag{2.21}$$

where $\mathbf{u}_x$ and $\mathbf{u}_z$ are unit vectors in the directions of the axes of x and z. The unit vector in the direction of propagation of the plane wave is

$$\mathbf{u} = \mathbf{u}_x \sin\theta + \mathbf{u}_z \cos\theta \tag{2.22}$$

Abbreviating this by writing

$$s = \sin\theta$$

and

$$c = \cos\theta = \sqrt{(1-s^2)} \tag{2.23}$$

the direction unit vector becomes

$$\mathbf{u} = \mathbf{u}_x s + \mathbf{u}_z c \tag{2.24}$$

in which s and c are the direction cosines. Hence the projection of OP on the direction of the plane wave is

$$\mathbf{u} \,.\, \mathbf{r} = sx + cz \tag{2.25}$$

If the vector electric field, as it passes the origin O, is $\mathbf{E}_\mathrm{o}$, the electric field at the point P will be

$$\mathbf{E}(x,z) = \mathbf{E}_\mathrm{o} \exp\{-jk\,(sx + cz)\} \tag{2.26}$$

(see equation (2.14)). It will be convenient to resolve the field into two orthogonal linearly polarized plane waves, one with the electric vector parallel to the x-z

plane and the other with the electric vector perpendicular to the plane. The first of these linearly polarized plane waves has its vector magnetic field pointing entirely in the transverse y-direction, and will therefore be referred to as the **transverse magnetic** (TM) case. The second has its electric field pointing entirely in the transverse direction, and will be referred to as the **transverse electric** (TE) case. We will treat the TM case first and the TE case, which is simply its dual, later.

2.2.1 Angular spectrum for transverse magnetic (TM) fields

All the plane waves in the angular spectrum which represents a transverse magnetic field in two diemensions will have their electric field lying in the plane of propagation, as depicted in Fig. 2.7. The magnetic field will then be wholly transverse.

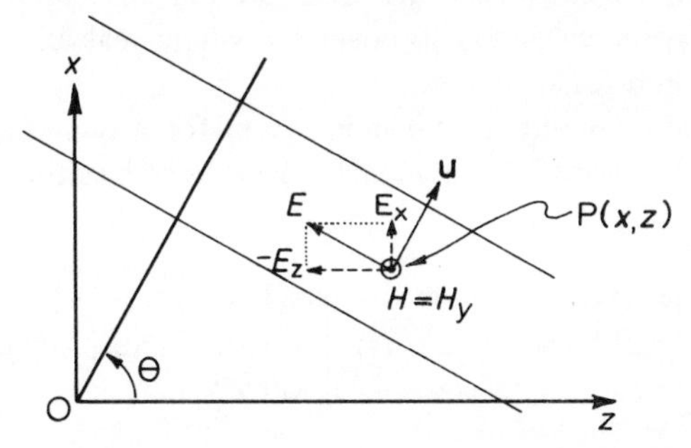

Fig. 2.7 – Showing the field components at point P(x,z) for a plane wave whose magnetic field is wholly transverse to the x-z plane.

If E_o is the phasor amplitude of the electric field of the plane wave as it passes the origin, then the Cartesian components of the electromagnetic field at P(x,z) will be

$$
\begin{aligned}
E_x(x,z) &= E_o \cos\theta \exp\{-jk(sx + cz)\} \\
E_z(x,z) &= -E_o \sin\theta \exp\{-jk(sx + cz)\} \\
H_y(x,z) &= Z^{-1}E_o \exp\{-jk(sx + cz)\}
\end{aligned}
\tag{2.27}
$$

with the remaining components, E_y, H_x and H_z all zero. Z is the characteristic impedance and k the phase constant of the medium.

A set of plane waves travelling in different directions is most conveniently represented by a spectrum function $A(\theta)$, such that the electric-field amplitude of the plane wave travelling in the direction θ is $A(\theta)\mathrm{d}\theta$. This spectrum function is completely analogous to the frequency spectrum function of time-series analysis, and like the frequency spectrum the angular spectrum $A(\theta)$ can be continuous, discrete, or a mixture of the two. The field components are obtained by replacing E_o by $A(\theta)\mathrm{d}\theta$ in the above equations, and then integrating over the range of angles for which the angular spectrum is defined. For example the x-component of the electric field will be

$$
E_x(x,z) = \int A(\theta) \cos\theta \exp\{-jk(sx + cz)\}\,\mathrm{d}\theta \tag{2.28}
$$

For reasons which will be explained in the next section, it is simpler to express the angular dependence of the spectrum in terms of $s = \sin\theta$, rather than in terms of θ itself. So replacing $A(\theta)$ by $F(s)$, and noting that $ds = \cos\theta \, d\theta$, equation (2.28) becomes

$$E_x(x,z) = \int F(s) \exp\{-jk(sx + cz)\} \, ds \tag{2.29}$$

which gives the x-component of the electric field at the point (x,z) as the integrated effect of all the plane waves in the angular spectrum $F(s)$. Note that $c = \cos\theta$ is retained as a convenient abbreviation for $\sqrt{(1-s^2)}$.

According to relation (2.20) the range of integration for s would be ± 1. However, it will be shown in the next section that for completeness the s-integration has to be extended to cover the whole real line, and hence that the limits of integration for s are $\pm\infty$.

The field components at the point (x,z) for a two-dimensional transverse magnetic field, in terms of the angular plane-wave spectrum function $F(s)$, are therefore

$$\begin{bmatrix} E_x(x,z) \\ E_z(x,z) \\ H_y(x,z) \end{bmatrix} = \int_{-\infty}^{\infty} F(s) \begin{bmatrix} 1 \\ -s/c \\ (Zc)^{-1} \end{bmatrix} \exp\{-jk(sx + cz)\} \, ds \tag{2.30}$$

in which $c = \sqrt{(1-s^2)}$. The important point to note at this stage is that the fields anywhere on and to the right of the aperture plane, that is, in the half-space $z \geqslant 0$, have been expressed in the TM case in terms of the single spectrum function $F(s)$. It will be shown later that the remaining field components, which are transverse electric, can be represented by a second spectrum function. But before doing so we will examine some of the important features of the representation of fields by an angular spectrum of plane waves by looking at equation (2.30) in rather more detail.

Exercise

Show by substituting into equation (2.30) that the discrete angular plane-wave spectrum $F(s) = E_o c_o \, \delta(s - s_o)$, where $s_o = \sin\theta_o$, $c_o = \cos\theta_o$ and $\delta(\;)$ is the Dirac delta function, represents a plane wave of amplitude E_o travelling in the direction $\theta = \theta_o$.

Exercise

Find the field components of the two-dimensional TM field whose angular spectrum is

$$F(s) = \frac{E_o c_o}{2} \left[\delta(s - s_o) + \delta(s + s_o) \right]$$

which is in fact the interference pattern of two inclined plane waves of equal amplitude. Sketch the planes of constant amplitude and constant phase. Show, for any of the three field components, that adjacent planes over which its amplitude is zero are separated by a distance $\lambda/(2s_o)$, and that planes of constant phase between which the phase differs by 2π are separated by λ/c_o, where $c_o = (1-s_o^2)^{1/2}$.

Exercise
The composite field examined in the previous exercise is of a type known as an **inhomogeneous plane wave**, since amplitude and phase are both constant over non-coincident planes. Deduce the propagation constant and wave impedance (defined as the ratio of the orthogonal electric and magnetic field transverse to the direction of propagation) of this inhomogeneous plane wave.

Exercise
Determine where, in the fields examined in the previous two exercises, two infinitely thin, perfectly conducting, parallel planes may be introduced without disturbing the fields. (See Appendix A.10, for a fuller discussion, from the point of view of TM modes in a parallel-plate waveguide.)

2.3 EVANESCENT WAVES

In order to get a better idea of the physical meaning of the plane-wave spectrum representation of the two-dimensional TM field of equation (2.30), consider the elemental contribution to the x-component of the electric field

$$\mathrm{d}E_x\,(x,z) = F(s)\,\mathrm{d}s\,\exp\{-jk\,(sx + cz)\} \tag{2.31}$$

which is the contribution of that plane wave in the spectrum of amplitude $F(s)$ ds travelling in the direction making an angle $\theta = \sin^{-1} s$ to the z axis.

When $|s| \leqslant 1$, θ lies in the range $-\pi/2 \leqslant \theta \leqslant \pi/2$, and the elemental plane wave of equation (2.31) is of the homogeneous type described in section 2.1. The wave travels with characteristic speed of the medium $(\mu\epsilon)^{-1/2}$, and transfers power into the half-space $z \geqslant 0$.

But when $|s| > 1$, assuming s still to be real, the character of the wave changes because the cosine

$$c = \sqrt{(1 - s^2)} = \pm j\chi \quad (\chi \text{ real and positive}) \tag{2.32}$$

is now imaginary. Substituting this into equation (2.31), and noting that we must choose $c = -j\chi$ in order that the fields remain finite as $z \rightarrow +\infty$,

$$\mathrm{d}E_x(x,z) = F(s)\,\mathrm{d}s\;\exp(-jksx)\exp(-k\chi z)\;, \tag{2.33}$$

This is a plane wave of inhomogeneous type, in that the amplitude is no longer constant over planes of constant phase. It has the following properties:

(a) The direction of propagation of the wave is along the x-axis, that is, parallel to the aperture plane, either positive or negative depending on the sign of s. The wavefronts over which the phase is constant are shown in Fig. 2.8.

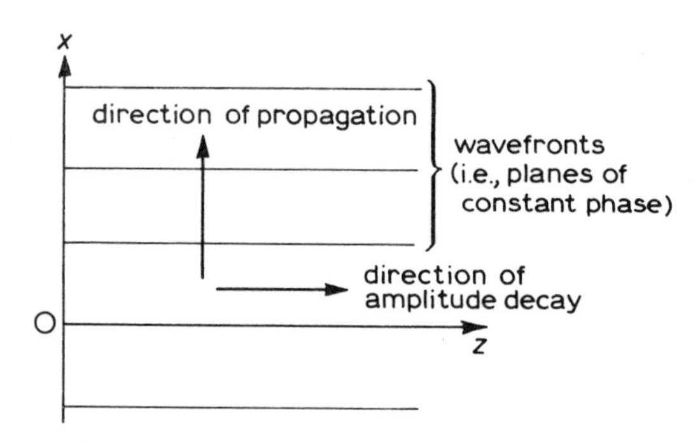

Fig. 2.8 – When $|s| > 1$ the plane waves of the angular spectrum $F(s)$ are inhomogeneous.

(b) The amplitudes of the field components decrease exponentially in the $+z$ direction, away from the aperture plane. For this reason they are often called **evanescent**, which means disappearing. The distance from the aperture plane at which the amplitude of the elemental plane wave of equation (2.33) has decreased to a fraction e^{-1} (= 0.3679) of its value at $z = 0$ is

$$z = \frac{\lambda}{2\pi\chi} \quad , \tag{2.34}$$

where χ is real and positive. For all but the smallest values of χ evanescent waves will have become negligibly small at distances of more than a few wavelengths from the aperture plane.

(c) The phase velocity of the evanescent wave, obtained by restoring the time dependence in equation (2.33), is

$$v_p = \frac{1}{s(\mu\epsilon)^{1/2}} \tag{2.35}$$

This velocity is slower than that of the homogeneous plane waves propagating in the medium, since $|s| > 1$. These slow waves do carry power, but it is not propagated into the region $z \geqslant 0$. It merely travels back and forth in the aperture plane, and can be thought of as being stored there. Because of their close association with a particular surface, in this case the aperture plane, these waves are also known as **electromagnetic surface waves.**

Exercise
Use the Poynting vector of equation (2.10) to determine the power flow in the elemental wave defined by equation (2.31) and its associated field components. Compare the two cases when $|s| < 1$ and $|s| > 1$.

Exercise
Define an aperture wave impedance for the elemental wave of equation (2.31) etc. as

$$Z_{ap} = \frac{dE_x}{dH_y} \tag{2.36}$$

which is the ratio of the transverse orthogonal components of the electric and magnetic field 'looking out' into the half-space $z \geqslant 0$. Show that Z_{ap} is imaginary when $|s| > 1$, and comment on the significance of this result.

Exercise
Apply the Poynting vector of equation (2.10) directly to the fields of equation (2.30) and show that the total power radiated into the half-space $z \geqslant 0$ is

$$P_{rad} = \frac{\lambda}{2Z} \int_{-1}^{+1} c^{-1} \, |F(s)|^2 \, ds \tag{2.37}$$

It may be useful in achieving this result to note that

$$\int_{-\infty}^{\infty} \exp \{\pm jkx \, (s - s')\} \, dx = \lambda \delta \, (s - s') \tag{2.38}$$

in which $k = 2\pi/\lambda$ and $\delta(\ )$ is the Dirac delta function.

Example
A very instructive physical example of the type of wave we have been discussing occurs on the far side of a dielectric boundary at which total internal reflection occurs. (See Fig. 2.9 and Appendix A.9). In this example the aperture plane

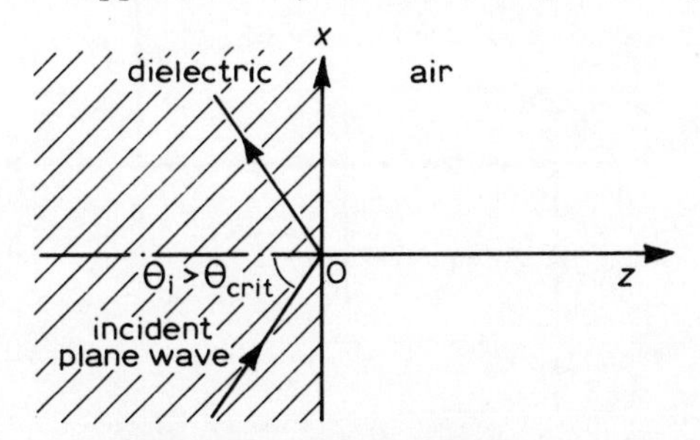

Fig. 2.9 – A plane wave incident at a dielectric/air boundary at an angle of incidence greater than critical.

will be taken to be coincident with the boundary between the dielectric and air. With a plane wave incident from the dielectric side at an angle of incidence θ_i greater than the critical ange θ_{crit} (for which the plane wave transmitted into the air medium would travel in a direction just parallel to the aperture plane) the incident wave will be totally reflected. However, the fields to the right of the aperture plane cannot be zero, since the boundary conditions (see Appendix B2.3) require that the tangential components of **E** and **H** be continuous across it. In fact the fields on the air side of the boundary are evanescent waves, which are local to the boundary and carry no energy across it. Thus writing Snell's law as

$$s = \sin\theta_t = (\epsilon_r)^{1/2}\sin\theta_i \tag{2.39}$$

in which θ_t is the angle of transmission corresponding to the angle of incidence θ_i, and ϵ_r is the relative permittivity of the dielectric, it is clear that when $\theta_i > \theta_{crit}$ $(= \sin^{-1}(\epsilon_r)^{-1/2})$ s becomes greater than unity but remains real. These are precisely the conditions for which we deduced the properties of evanescent waves.

Mathematical Comment

What is the nature of θ when the magnitude of $\sin\theta$ is greater than unity? The answer is that θ is in general a complex angle. If we apply the two conditions:

(I) s is real and in the range $-\infty < s < \infty$

(II) when $|s| > 1$ $c = \sqrt{(1 - s^2)} = -j\chi$ is negative imaginary, that is, χ is real and > 0, (2.40)

we can deduce that the contour Γ of θ in the complex-θ plane, corresponding to s traversing the real line from $-\infty$ to $+\infty$, is that given in Fig. 2.10. That part of Γ for which θ is real, namely $-\pi/2 \leqslant \theta \leqslant \pi/2$, corresponds to $-1 \leqslant s \leqslant 1$.

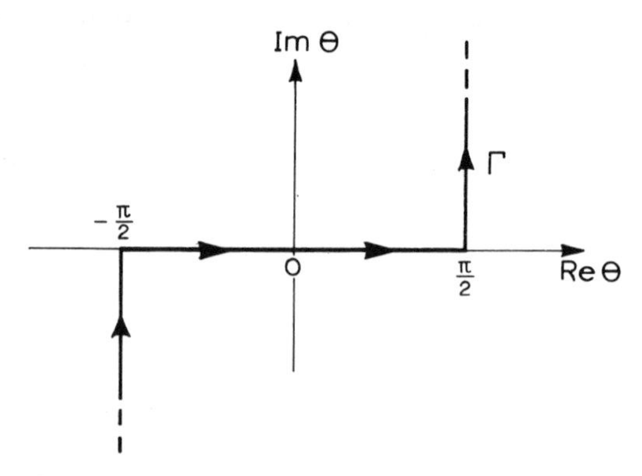

Fig. 2.10 – Representation of θ in the complex-θ plane: contour Γ corresponds to $s = \sin\theta$ traversing the real line from $-\infty$ to $+\infty$.

The remainder of the contour Γ goes off to infinity in directions which ensure that the fields in the angular plane-wave spectrum representation of equation (2.30) are bounded as $z \to \infty$.

Exercise
By writing $\theta = \theta' + j\theta''$, where θ' and θ'' are the real and imaginary parts of θ, and applying conditions I and II above, show that the contour Γ is indeed that shown in Fig. 2.10.

Exercise
Rewrite the field representation of equation (2.30) as an integral over θ rather than over $s = \sin\theta$.

Summarising the discussion of this section, a complete representation of the two-dimensional fields in a half-space $z \geqslant 0$ in terms of an angular spectrum of plane waves $F(s)$, in which $s = \sin\theta$, must include spectral components over the entire range $-\infty < s < +\infty$. Transitions in the nature of the field occur at $s = \pm 1$. Over the sector $|s| \leqslant 1$ the plane waves propagate freely as homogeneous plane waves into the region $z \geqslant 0$. This is known as the **visible** or **propagating** part of the angular spectrum. Over the remainder of the range of s, namely $|s| > 1$, the plane waves are inhomogeneous, do not propagate into the medium but only store reactive power in the aperture plane, and are evanescent in that they are negligible at more than a few wavelengths distant from the aperture plane. This is the **invisible, non-propagating** or **reactive** part of the angular spectrum. Its inclusion has important practical consequences. But for our immediate purposes the most significant thing about being able to extend s over the whole real line is that this formulation lends itself to being couched in terms of Fourier transforms. In particular it leads to the following fundamental Fourier-transform relationship between the aperture field and the angular plane-wave spectrum.

2.4 APERTURE FIELD : FOURIER TRANSFORM OF THE ANGULAR SPECTRUM

Consider the x-component of the electric field over the aperture plane $z = 0$, and denote it by the additional suffix 'a'. Thus

$$E_{ax}(x) = E_x(x,0) \tag{2.41}$$

and for the TM field of equation (2.30)

$$E_{ax}(x) = \int_{-\infty}^{\infty} F(s) \exp(-jksx)\, ds \tag{2.42}$$

which states that the aperture field $E_{ax}(x)$ is the Fourier transform of the angular plane-wave spectrum $F(s)$. The inversion formula for Fourier transforms yields the angular spectrum as

$$F(s) = \frac{1}{\lambda} \int_{-\infty}^{\infty} E_{ax}(x) \exp(+jksx) \, dx \tag{2.43}$$

This means that if we know the tangential component of the electric field over the aperture plane $z = 0$, we can deduce by means of equation (2.43) the angular spectrum, in terms of which we know the fields everywhere in the region $z \geqslant 0$, from equation (2.30). This statement is *exact,* and applies to transverse-magnetic fields in two dimensions.

We may on occasion write the Fourier-transform relationship between $F(s)$ and $E_{ax}(x)$ symbolically as

$$F(s) \longleftrightarrow E_{ax}(x) \tag{2.44}$$

or even denote the scalar component $E_{ax}(x)$ by $f(x)$ so that

$$F(s) \longleftrightarrow f(x) \tag{2.45}$$

which is not only briefer but also emphasizes the analogy with the corresponding relationship in circuit theory

$$F(\omega) \longleftrightarrow f(t) \tag{2.46}$$

for the frequency spectrum $F(\omega)$ of the time waveform $f(t)$, given explicitly by equations (1.4) and (1.5).

Example of diffraction of a plane wave by a slit of width a in a thin, perfectly conducting plane screen

Suppose that the screen is coincident with the x-y plane and that the slit is parallel to, and symmetrical about, the y-axis, as depicted in Fig. 2.11. If the plane wave is incident normally on the screen and is linearly polarized with its electric field wholly x-directed, that is,

$$E_x^{\text{inc}}(x,z) = E_0 \exp(-jkz) \quad , \tag{2.47}$$

the diffracted field will be a two-dimensional field of the TM type, and the formulas previously developed will apply.

The appropriate aperture field is the x-component of the electric field just to the right of the screen, that is, at $z = 0+$. The tangential electric field on the

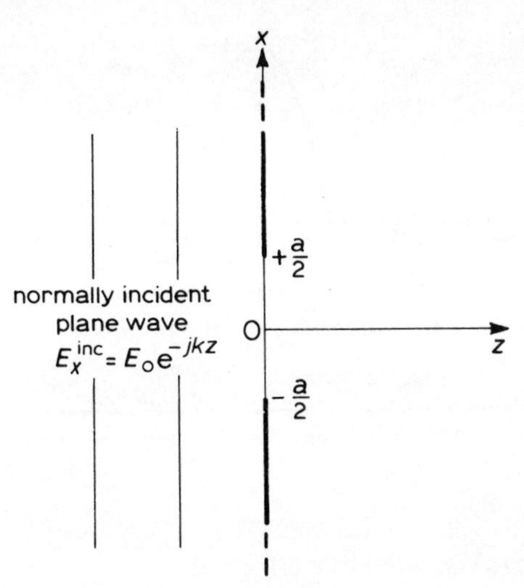

Fig. 2.11 – A slit in a thin, perfectly conducting screen illuminated by a normally incident plane wave.

conducting screen must be zero, in order that the boundary conditions (Appendix B 2.3) are satisfied. Across the opening of the slit the field can be taken as continuous with the incident field at $z = 0-$. Hence the aperture field is approximately

$$E_{ax}(x) = E_0 \text{ for } |x| < a/2$$

$$= 0 \text{ otherwise.} \tag{2.48}$$

The angular spectrum of the field diffracted into the region $z \geqslant 0$, by substitution into equation (2.43),

$$F(s) = \frac{E_0}{\lambda}\int_{-a/2}^{a/2} \exp(jksx)\, dx$$

that is,

$$F(s) = E_0 \frac{a}{\lambda} \frac{\sin\left(\dfrac{\pi as}{\lambda}\right)}{\dfrac{\pi as}{\lambda}} = E_0 \frac{a}{\lambda} \text{ sinc}\left(\frac{as}{\lambda}\right) \tag{2.49}$$

The 'sinc' function employed in equation (2.49) is the real symmetrical function

$$\text{sinc}(u) = \frac{\sin(\pi u)}{\pi u} \tag{2.50}$$

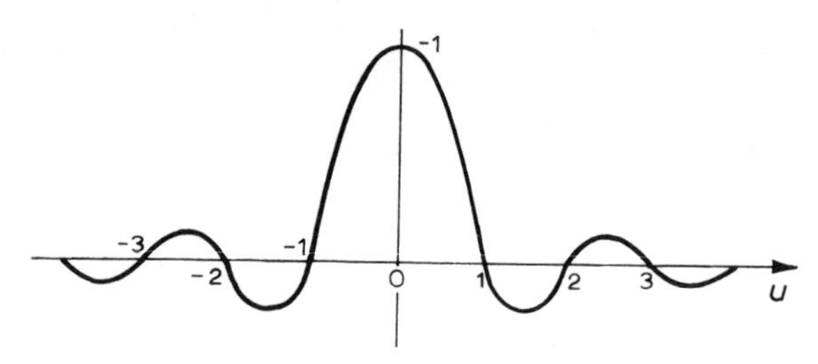

Fig. 2.12 – Graph of sinc $(u) = \sin(\pi u)/(\pi u)$.

which is sketched in Fig. 2.12. Some of its properties are that

$$\text{sinc}\,(0) = 1 \quad , \tag{2.51}$$

$$\int_{-\infty}^{\infty} \text{sinc}\,(u)\, du = 1 \quad , \tag{2.52}$$

its zeros occur at $u = \pm\, 1, \pm\, 2$, etc., (2.53)

the minimax values between successive zeros are approximately

$$-\frac{2}{3\pi}\, ,\; +\frac{2}{5\pi}\, ,\; -\frac{2}{7\pi}\, ,\; \text{etc.} \tag{2.54}$$

Applying these properties to the diffracted angular spectrum of equation (2.49), it has the same shape as the graph of Fig. 2.12, with

$$u = \frac{as}{\lambda} = \frac{a \sin\theta}{\lambda} \tag{2.55}$$

The absolute maximum of the spectrum occurs in the direction $\theta = 0$, that is, in the direction of the z-axis. Its amplitude in that direction is

$$F(0) = \frac{E_0 a}{\lambda} \quad , \tag{2.56}$$

which increases as a/λ increases. In the limit, as $a/\lambda \to \infty$, the angular spectrum is given by

$$F(s) = E_0 \lim_{a/\lambda \to \infty} (a/\lambda)\, \text{sinc}\,(as/\lambda) = E_0 \delta(s) \tag{2.57}$$

which is a plane wave travelling in the direction $\theta = 0$, as it should be.

The first zeros of the diffraction pattern occur at $\theta = \pm\,\theta_0$, where

$$\sin\theta_0 = \lambda/a \quad . \tag{2.58}$$

The beamwidth between these first zeros is then

$$2\theta_0 = 2\sin^{-1}(\lambda/a) \tag{2.59}$$

which decreases as a/λ increases. The level of the first sidelobes of the pattern is −13.5 dB with respect to the main-lobe maximum, and is independent of a/λ. It will be shown later that the *shape* of the aperture field distribution determines the sidelobe level.

Exercise
Prove equation (2.57).

Plane-wave components in the angular spectrum are evanescent when $|s| > 1$; which corresponds, from equation (2.55), to when $|u| > a/\lambda$. The diffraction patterns shown in Fig. 2.13 contrast the two extremes of when the slit is very many wavelengths wide and when it is only a small fraction of a wavelength wide. When $a \gg \lambda$ the amplitude of the spectrum will be very small in the evanescent region; whereas when $a \ll \lambda$ the spectrum is essentially constant over the propagating range of angles, and continues at approximately the same level far into the evanescent region.

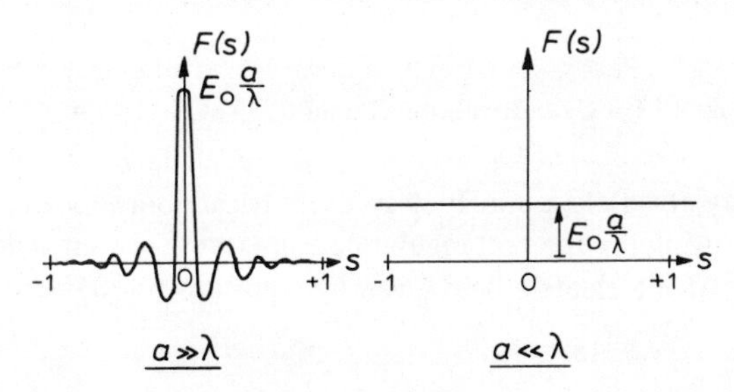

Fig. 2.13 – Diffraction patterns for a slit that is very wide and one that is very narrow, in terms of the wavelength.

It has been suggested above, for example in Fig. 2.13, that the angular spectrum and the diffraction pattern are equivalent. The fact is that at large distances from the aperture the angular dependence of the fields is approximately that of the angular spectrum, an approximation that improves the larger the distance from the aperture. The details of this important relationship between the far field (as it is known in antenna practice, or the Fraunhofer field, as it is known in optics) and the angular plane-wave spctrum will be given in the next section.

2.5 FAR FIELD : APPROXIMATED BY THE ANGULAR SPECTRUM

Any two-dimensional, transverse-magnetic field propagating into the region $z \geqslant 0$ can be represented by the single spectrum function $F(s)$ as in equation (2.30). Take for example the x-component of the electric field:

$$E_x(x,z) = \int_{-\infty}^{\infty} F(s) \exp\{-jk(sx + cz)\} \, ds \tag{2.60}$$

In principle E_x is determined everywhere in $z \geqslant 0$ by a single integral; but in practice the integral can be difficult to evaluate. However, there is one important general result that we can obtain immediately for the field at very large distances from a diffracting aperture of finite size. This result will be obtained by evaluating the integral (2.60) by the 'method of stationary phase', which will first be described in physical terms and then developed mathematically. Finally, an equivalent but rather more rigorous technique for evaluating the integral, known as the 'method of steepest descent', will be given.

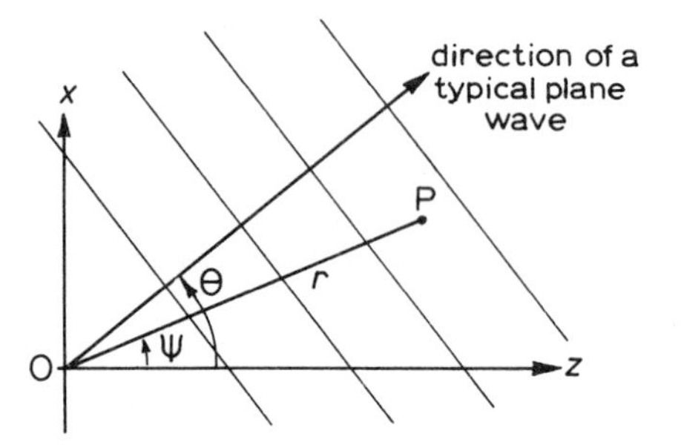

Fig. 2.14 – Cylindrical coordinates (r, ψ) of field point P.

First, identify the field point P by its cylindrical coordinates (r, ψ) shown in Fig. 2.14. Substituting for the rectangular coordinates $x = r \sin \psi$ and $z = r \cos \psi$, the x-component of the electric field given by equation (2.60) becomes

$$\begin{aligned} E_x(r, \psi) &= \int_{-\infty}^{\infty} F(s) \exp\{-jkr(s \sin \psi + c \cos \psi)\} \, ds \\ &= \int_{-\infty}^{\infty} F(\sin \theta) \exp\{-jkr \cos(\psi - \theta)\} \, d(\sin \theta) \end{aligned} \tag{2.61}$$

Now suppose that the point P is many wavelengths distant from the point of origin O, that is, $kr \gg 1$. For most of the range of ψ evanescent waves will not contribute. As θ varies over the range of real angles the phase in the exponential term of the integrand in equation (2.61) will rotate rapidly through many multiples of 2π, *except* when $\cos(\psi - \theta)$ is stationary. The stationary-phase condition occurs when $\theta = \psi$, that is, for that wave in the angular spectrum travelling in the direction OP.

Next suppose that $F(s)$ is a bounded and continuous function of s, which means, for example, that it must not contain any discrete delta-function components. This condition implies that the aperture field can be non-zero only over a finite area of the aperture plane, which is a perfectly acceptable condition in practice. Then as angle θ, and hence $\sin\theta$, varies over the integration range, neighbouring values of $F(s)$ can occur which have approximately the same amplitude but appear in the integrand of equation (2.61) with opposite phases, hence almost cancelling each other. Pairs of values of the integrand tending to cancel one another can be expected to occur throughout the entire range of integration, *except* in the direction in which the stationary-phase condition is satisfied. The only non-negligible contribution to the integrand of equation (2.61) will therefore come from the direction $\theta = \psi$ and its immediate neighbourhood, and we may write the x-component of the electric field at P as

$$E_x(r, \psi) \simeq CF(\sin\psi)\exp(-jkr) \tag{2.62}$$

in which C is a constant soon to be determined, and the sign $\simeq$ is to be read as 'asymptotically equal to' since it is clear that the foregoing physical argument becomes more exact as $kr \to \infty$, at which point we expect equation (2.62) to become an equality.

The **method of stationary phase** applied to an integral of the form

$$I = \int_a^b f(x)\exp\{jKg(x)\}\,\mathrm{d}x \tag{2.63}$$

in which $f(x)$ and $g(x)$ are real, continuous functions of bounded variation, and K is a large positive real number, will now be given in detail in order that the constant C in equation (2.62) may be determined.

Suppose that the phase function $g(x)$ is stationary at only the one point $x = x_0$ in the range of integration, that is,

$$g'(x_0) = 0 \ ; \ a < x_0 < b \ , \tag{2.64}$$

where the prime on the function g means taking its derivative.

According to the physical argument given above, the only significant contribution to the integral of equation (2.63) will be from the neighbourhood of the stationary point. Let $x = x_0 + \xi$, and expand $g(x)$ in a Taylor series about the point $x = x_0$, that is,

$$g(x) = g(x_0) + 0 + \tfrac{1}{2}\xi^2 g''(x_0) + \ldots \tag{2.65}$$

and assume that $g''(x_0) \neq 0$. Then, assuming that $f(x)$ is only slowly varying, we may write the integral of equation (2.63) approximately as

$$I \simeq f(x_0) \exp\{jkg(x_0)\} \int_{-\infty}^{\infty} \exp\{j\tfrac{1}{2}Kg''(x_0)\xi^2\}\, d\xi \tag{2.66}$$

where the limits of integration can now be extended to $\pm\infty$ without affecting the argument. Equation (2.66) is an asymptotic result since we expect the approximation to improve as $K \to \infty$; an expectation that is borne out by the more rigorous analysis given below based on the method of steepest descent.

The integral in equation (2.66) is now of the standard type:

$$\int_{-\infty}^{\infty} \exp(-\alpha\xi^2)\, d\xi = \sqrt{(\pi/\alpha)} \text{ for } \operatorname{Re} \alpha > 0 \tag{2.67}$$

with

$$\alpha = -j\tfrac{1}{2}K|g''(x_0)| \operatorname{sgn}[g''(x_0)] \tag{2.68}$$

where sgn( ) is the signum function. Hence if we may assume that α has a very small positive real part, equation (2.66) becomes

$$I \simeq \sqrt{\frac{2\pi}{K|g''(x_0)|}}\, f(x_0) \exp\{j\,(Kg(x_0) - (\pi/4) \operatorname{sgn}[g''(x_0)]\,)\} \tag{2.69}$$

This result will be referred to as the stationary-phase algorithm for the asymptotic evaluation, as $K \to \infty$, of single integrals of the type of equation (2.63).

Applying the stationary-phase algorithm to the diffraction integral of equation (2.61), we can identify the integration variable x with θ, the stationary-phase point x_0 with ψ, K with kr, $f(x)$ with $\cos\theta\; F(\sin\theta)$, and $g(x)$ with $-\cos(\theta\text{-}\psi)$. Hence for $kr \to \infty$

$$E_x(r, \psi) \simeq \sqrt{\frac{j\lambda}{r}} \cos\psi\; F(\sin\psi) \exp(-jkr) \tag{2.70}$$

which is the completed version of equation (2.62). The remaining components of the far field can be similarly derived, and are

$$E_z(r, \psi) \simeq -\sqrt{\frac{j\lambda}{r}} \sin\psi\; F(\sin\psi) \exp(-jkr) \tag{2.71}$$

and

$$H_y(r,\psi) \simeq \frac{1}{Z}\sqrt{\frac{j\lambda}{r}}\, F(\sin\psi)\exp(-jkr) \tag{2.72}$$

The **method of steepest descent** provides a more rigorous justification of the stationary-phase algorithm of equation (2.69) than the heuristic arguments given above. It also furnishes proof that equation (2.69) is the first term of an asymptotic series in ascending odd powers of $K^{-1/2}$, and hence that (2.69) is indeed an asymptotic result. We will give only an outline of the method here. The general method of evaluation of integrals by steepest descent can be found in Dennery and Krzywicki (1967), and the particular application to the evaluation of the far field in plane-wave spectrum theory is given in Clemmow (1966).

Equation (2.61) can be expressed as a contour integral, along the contour Γ in Fig. 2.10, as

$$E_x(r,\psi) = \int_\Gamma \cos\theta\, F(\sin\theta)\exp\{-jkr\cos(\theta-\psi)\}\, d\theta \tag{2.73}$$

Along the real θ axis the argument of the exponential is purely imaginary, and is a minimum at the point $\theta = \psi$. If we now suppose that the argument of the exponential is in general the complex analytic function $z(\theta)$, we can view the real axis as the direction along which Re z = constant = 0, and Im z reaches a minimum. But we know from the general theory of a complex variable that neither Re z nor Im z can have an absolute maximum or minimum, since z is analytic. Hence the point $\theta = \psi$ must be a saddle point, for both Re z and Im z; which is why the method of steepest descent is often referred to as the **saddle-point method.** But more important, we can find a path through the point $\theta = \psi$ along which the real part of z is a maximum at the saddle point and descends rapidly on either side to large negative values. This is the path of steepest descent for Re z and happens to coincide with Im z = constant. The complete descent path Γ_s is shown in Fig. 2.15, in comparison with the original integration contour Γ.

Since the integrand contains no singularities, Cauchy's theorem assures us that Γ_s can equally well be taken as the path along which to evaluate the integral. Then, because the argument of the exponential in the integrand of equation (2.73) is real and decreases rapidly to large negative values on either side of the saddle point, the highly localized contribution of the integrand to the integral is confirmed; increasingly so as $kr \to \infty$.

Now returning to the components of the far field, given by equations (2.70) to (2.72) and for the sake of economy in symbols replacing the polar angle ψ of the

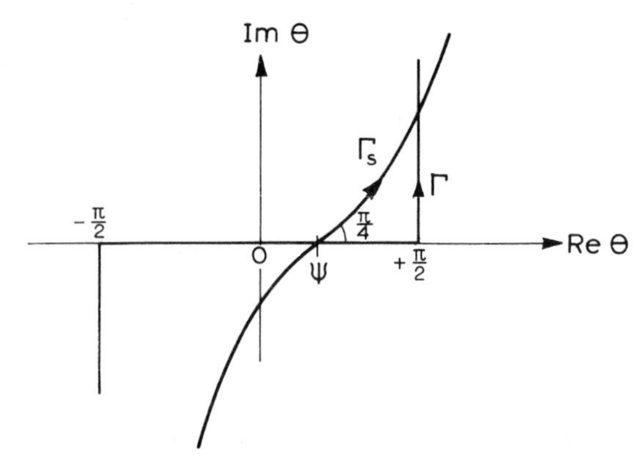

Fig. 2.15 – The contour Γ_s is the path of steepest descent through $\theta = \psi$.

field point P by θ, it is seen with the aid of Fig. 2.16 that the far field for a TM polarized radiating field in two dimensions is a cylindrical wave of the asymptotic form ($kr \to \infty$):

$$E_\theta(r,\theta) = Z\,H_y(r,\theta)$$

$$= \sqrt{\frac{\lambda}{r}}\;F(\sin\theta)\exp\{-j\,[kr - \pi/4]\} \tag{2.77}$$

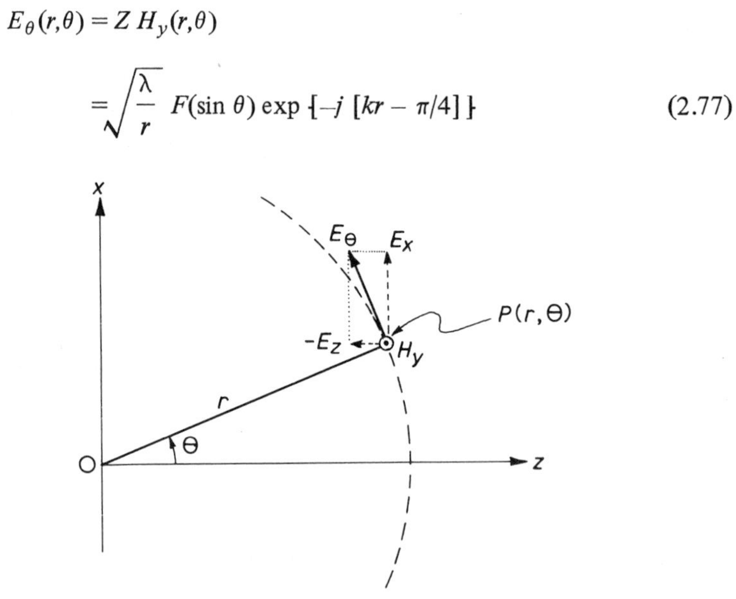

Fig. 2.16 – TM components of the far field at the point (r,θ). (Note change of angle designation from ψ to θ.)

A wave of this type is called cylindrical because the phase is constant over cylindrical surfaces, in this case at constant r with the cylinder axis coinciding with the y-axis. The amplitude of the wave decreases in any radial direction as $r^{-1/2}$, and its angular dependence is given by the angular spectrum $F(\sin\theta)$.

The character of the field local to any point P in the far field is approximately that of a plane wave, as is evident from equation (2.77). (From a purely physical point of view, the cylindrical wavefront becomes increasingly plane as r tends to infinity). Poynting's theorem applied to equation (2.77) shows that the power flow in the far field is radial at all points and is of magnitude

$$S_r(r,\theta) = \frac{\lambda}{2Zr} \; |F(\sin\theta)|^2 \tag{2.78}$$

Thus the radial power flow, which has the dimensions of watts per metre per unit width, falls off in any radial direction as r^{-1}. Its angular dependence is given by the squared modulus of the angular spectrum, namely $|F(\sin\theta)|^2$, which is therefore often referred to as the **angular power spectrum.**

Antenna Far-Field Theorem

It is worth drawing attention here to the main results of this section by stating them in the form of a theorem:

> The far field of an antenna is asymptotically equal to its angular spectrum. The angular spectrum for an antenna is the Fourier transform of its aperture field, which is the tangential component of the field in the aperture plane of the antenna.

We have only proved this result for one of the orthogonally polarized components of a two-dimensional radiating field, in equations (2.77) and (2.43). However, it will be shown in the next chapter that the above theorem is true in the three-dimensional case. It is common in antenna practice (see section 1.4) to define the far field over some sphere of sufficiently large radius. In two dimensions the sphere becomes a circle. The word 'asymptotically' used in the theorem implies that $kr \to \infty$, where r is the radius of the sphere or circle of observation. In practice, as will be shown in section 5.3.1, it is usually satisfactory for most purposes to ensure that

$$r \geqslant \frac{2a^2}{\lambda} \tag{2.79}$$

where a is the dimension of the aperture for a one-dimensional aperture field, or is the larger dimension in the case of a two-dimensional aperture field. It is usual to choose the tangential component of the electric field to define the aperture field, but it is equally valid to choose the tangential component of the magnetic field, which is why it is left unspecified in the theorem. It is no mere pedantry that prevents us from stating the above theorem in a more abbreviated, and perhaps more familar, form such as 'The far field of an antenna is given approximately by the Fourier transform of its aperture field'. It is important to keep both statements in the theorem separate in order to emphasize that the Fourier transform relationship between angular spectrum and aperture field is exact, and that the approximation enters in the evaluation of the far field.

Exercise
Sketch the amplitude and phase of the far field pattern arising from a one-dimensional aperture field distribution which is constant in both amplitude and phase over a width of four wavelengths and zero over the remainder of the aperture plane. At what distance from the centre of the aperture plane could one expect to measure this pattern when the wavelength is 0.1 m?

Exercise
Re-derive equation (2.37), which gives the total power radiated into the half-space $z \geqslant 0$, by integrating the power flow in the far field given by equation (2.78) over a cylindrical surface in the right-hand half-space.

Exercise
Form the self-convolution function

$$C_a(\xi) = \int_{-\infty}^{\infty} E_{ax}(x+\xi) E_{ax}^*(x)\, dx \tag{2.80}$$

of the aperture-field function $E_{ax}(x)$. Show that $C_a(\xi)$ is the Fourier transform of the angular power spectrum $|F(s)|^2$, given that $E_{ax}(x)$ is the Fourier transform of the angular spectrum $F(s)$.

Exercise
If $E_{px}(x)$ is the tangential electric field (in a two-dimensional TM situation) over some arbitrary plane z = constant, parallel to the aperture plane, and its self-convolution function is defined by

$$C_p(\xi) = \int_{-\infty}^{\infty} E_{px}(x+\xi) E_{px}^*(x)\, dx \quad ,$$

show that $C_p(\xi)$ is identical with $C_a(\xi)$ defined in the previous exercise. (This is the deterministic form of an important result in the theory of partial coherence, which states that the lateral coherence function of the field (in this case $C_p(\xi)$) travels without change in a uniform medium. The Fourier transform relation derived in the previous exercise is the van Cittert-Zernike theorem for deterministic fields.)

Geometrical Optics and the Far Field

A useful feature of the far field is that it obeys the laws of geometrical optics. In a uniform medium these laws are simply that rays travel in straight lines (rays being lines drawn orthogonally to the wavefront such that they define the local direction of propagation of the wave) and that the power flows along conical tubes of contiguous rays such that the power traversing the cross-section of any single tube is constant along the entire length of that tube. The rays associated with the far field of Fig. 2.16 are clearly radial lines drawn outward from O in the plane of the figure. Three such rays are drawn in Fig. 2.17, the

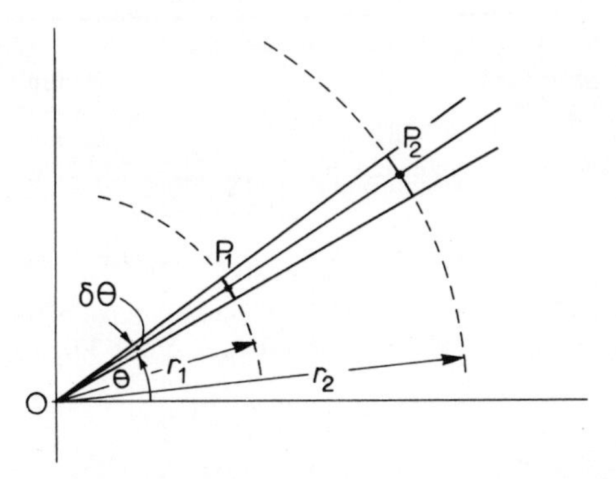

Fig. 2.17 – The far field depicted as a wedge of rays passing through the field points P_1 and P_2.

central ray passing through the field points P_1 and P_2 which are at distances r_1 and r_2 respectively from the origin O in some direction θ. Consider the ray tube to be the triangular wedge formed by the two outer rays and extending with unit width at right angles to the plane of the figure. For an infinitesimally small wedge angle $d\theta$, if the power traversing the cross-sectional surfaces of the wedge at P_1 and P_2 is to be the same,

$$S_r(r_1,\theta)r_1\,d\theta = S_r(r_2,\theta)r_2\,d\theta \tag{2.81}$$

which is obviously true for the far-field radial power flow of equation (2.78). The same result is true even if the wedge angle $\delta\theta$ is finite, as can be seen by integrating both sides of equation (2.81). However, the differential form of equation (2.81) is often useful for deducing the magnitude of the complex amplitude of the field at P_2, say, when one knows the field at P_1, the phase difference being proportional to the ray path length P_1P_2, namely $k(r_2-r_1)$. The underlying technique of the geometrical optics method is the same even when the direction of the rays is changed by the intervention of lenses or reflecting surfaces, as will be explained in more detail in later chapters.

Fraunhofer Diffraction

In optics the far field pattern is referred to as the **Fraunhofer diffraction pattern,** the only difference being the minor one that the fields are commonly specified over a plane rather than over a sphere or circle (see Fig. 2.18). If the extent of

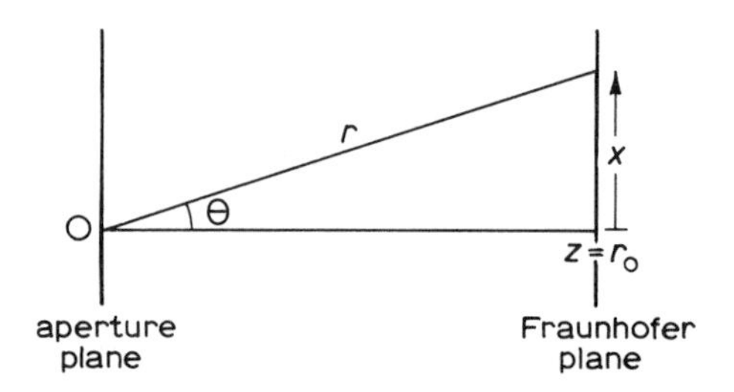

Fig. 2.18 – Geometry for Fraunhofer diffraction.

the observing (Fraunhofer) plane is very small compared with its distance r_o from the aperture plane, $\sin\theta$ can be approximated by x/r_o which is small, and the significant field components in the Fraunhofer plane are approximately (from equation (2.77))

$$E_x(x,r_o) = ZH_y(x,r_o)$$

$$= \sqrt{\frac{j\lambda}{r_o}} \exp\{-jk\left(r_o + \frac{x^2}{2r_o}\right)\} F\left(\frac{x}{r_o}\right) \tag{2.82}$$

in which the approximation $r = r_o + x^2/(2r_o)$ has been used in the exponential term.

Exercise

Write down the intensity (which is the term used in optics for power flow) of the field of equation (2.82) over the Fraunhofer plane.

Huygens' Principle

In two dimensions, Huygens' principle states that every point on a wavefront can be replaced by a secondary source radiating a cylindrical wave in the forward direction of the original wave. Consider the arbitrary wavefront shown in Fig. 2.19 and divide it into infinitesimal segments. Take the segment at O in isolation from the rest and consider it as radiating into the half-space $z \geqslant 0$. Then if we suppose that the wavefront is tangential to the x-axis at O, and that the electric field is wholly x-directed and of magnitude E at this point, the angular spectrum of the field radiated into the half-space $z \geqslant 0$ due only to the element at O is

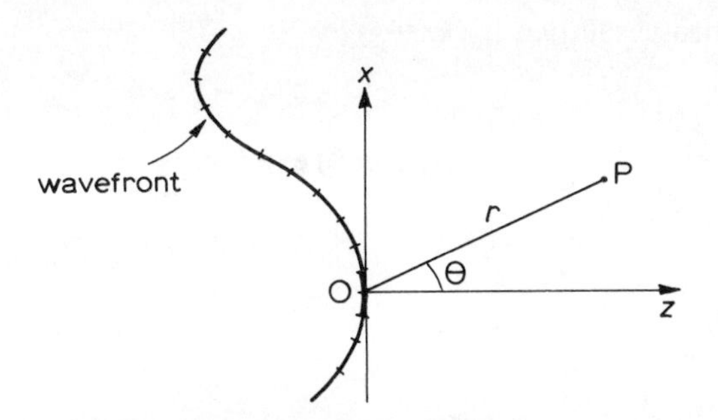

Fig. 2.19 – A small segment of an arbitrary wavefront at O considered as radiating into the half-space $z \geqslant 0$ establishes Huygens' principle.

$E\mathrm{d}x/\lambda$, where $\mathrm{d}x$ is the length of the element. (Substitute $\mathrm{d}x$ for a in equation (2.49)). Then equation (2.77) gives the elemental field at P as

$$\mathrm{d}E_\theta = Z\,\mathrm{d}H_y = E\,\mathrm{d}x\sqrt{\frac{j}{\lambda r}}\;\mathrm{e}^{-jkr} \tag{2.83}$$

which is indeed a uniform cylindrical wave emanating from O, as Huygens' principle prescribes. Equation (2.83) amplifies Huygens' principle by giving the precise strength of the secondary sources and the polarization of the field radiated by them. Note also that equation (2.79) guarantees that the point P will always be in the far field of the radiator of vanishingly short length.

Exercise

Use Huygens' principle to find the field at a point situated at a very large distance from an aperture field which is x-directed, of constant amplitude and phase over a width a, and zero elsewhere.

2.6 THE COMPLETE TWO-DIMENSIONAL FIELD : UNIQUENESS

The transverse-magnetic, two-dimensional field of equation (2.30) gives three, out of a possible total of six, field components in terms of a single angular spectrum which will now be designated $F_{\mathrm{TM}}(s)$. Thus

$$\begin{bmatrix} E_x(x,z) \\ E_z(x,z) \\ H_y(x,z) \end{bmatrix} = \int_{-\infty}^{\infty} F_{\mathrm{TM}}(s) \begin{bmatrix} 1 \\ -s/c \\ (Zc)^{-1} \end{bmatrix} \exp\{-jk(sx + zc)\}\,\mathrm{d}s \tag{2.84}$$

which has the asymptotic solution for $kr \to \infty$ of

$$\begin{bmatrix} E_x(r,\theta) \\ E_z(r,\theta) \\ H_y(r,\theta) \end{bmatrix} = \begin{bmatrix} \cos\theta \\ -\sin\theta \\ Z^{-1} \end{bmatrix} \sqrt{\frac{j\lambda}{r}}\, F_{TM}(\sin\theta) \exp(-jkr) \tag{2.85}$$

or, in polar form, of

$$E_\theta(r,\theta) = ZH_y(r,\theta) = \sqrt{\frac{\lambda}{r}}\, F_{TM}(\sin\theta) \exp\{-j(kr - \pi/4)\} \tag{2.86}$$

The transverse-electric (TE) field in two dimensions, which prescribes the three remaining field components, can be constructed in exactly the same way in terms of a second angular spectrum $F_{TE}(s)$. Rather than repeat the construction that led up to equation (2.30), it is simpler to invoke the principle of duality (see Appendix B 2.7) which enables us, literally, to write down the TE fields by inspection of the TM fields of equation (2.30) or its equivalent, equation (2.84).

The principle of duality applies in a source-free, lossless medium, and it states that if $(\mathbf{E}_1, \mathbf{H}_1)$ is a solution of Maxwell's equations in this medium then a second, independent solution $(\mathbf{E}_2, \mathbf{H}_2)$ can be obtained by making the substitutions $\mathbf{E}_2 = Z\mathbf{H}_1$ and $\mathbf{H}_2 = -\mathbf{E}_1/Z$. It is important to note that different boundary conditions must apply in order that the two solutions are truly independent. Thus

$$\begin{bmatrix} E_y(x,z) \\ H_x(x,z) \\ H_z(x,z) \end{bmatrix} = \int_{-\infty}^{\infty} F_{TE}(s) \begin{bmatrix} 1 \\ -c/Z \\ s/Z \end{bmatrix} \exp\{-jk(sx + cz)\}\, ds \tag{2.87}$$

where we have replaced $c^{-1} F_{TM}(s)$ by $F_{TE}(s)$. The aperture field will again be taken as the tangential electric field over the aperture plane, which in this case will be

$$E_{ay}(x) = E_y(x,0) \tag{2.88}$$

It is clear from equation (2.87) that $E_{ay}(x)$ is the Fourier transform of $F_{TE}(s)$, namely,

$$E_{ay}(x) = \int_{-\infty}^{\infty} F_{TE}(s) \exp(-jksx)\, ds \tag{2.89}$$

whose inverse yields the transverse-electric angular spectrum as being

$$F_{\mathrm{TE}}(s) = \frac{1}{\lambda} \int_{-\infty}^{\infty} E_{ay}(x) \exp(+jksx)\, \mathrm{d}x \qquad (2.90)$$

The solution of equation (2.87) in the far field is, again by applying the principle of duality to equation (2.86), asymptotically for $kr \rightarrow \infty$

$$E_y(r,\theta) = -ZH_\theta(r,\theta)$$

$$= \sqrt{\frac{\lambda}{r}}\, F_{\mathrm{TE}}(\sin\theta) \cos\theta \exp\{-j(kr - \pi/4)\} \qquad (2.91)$$

which is the cylindrical wave depicted in Fig. 2.20.

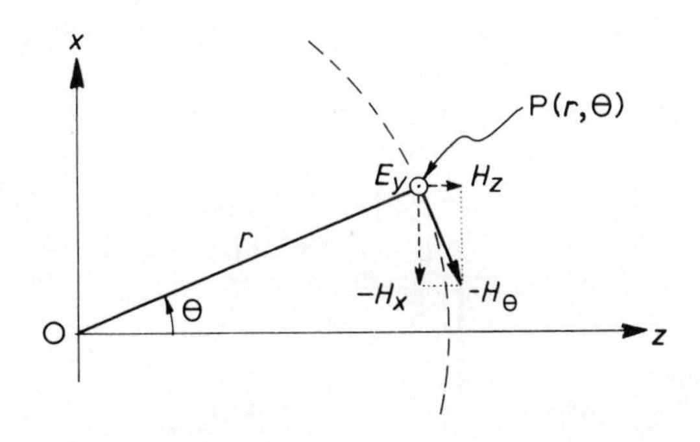

Fig. 2.20 – TE components of the far field at the point (r,θ).

In summary, the complete electromagnetic field, in two dimensions, radiated into the half-space $z \geqslant 0$ is given by equations (2.84) and (2.87) in terms of the two independent angular spectra

$$\begin{bmatrix} F_{\mathrm{TM}}(s) \\ F_{\mathrm{TE}}(s) \end{bmatrix} = \frac{1}{\lambda} \int_{-\infty}^{\infty} \begin{bmatrix} E_{ax}(x) \\ E_{ay}(x) \end{bmatrix} \exp(+jksx)\, \mathrm{d}x \qquad (2.92)$$

which are themselves defined by two orthogonal components of the tangential electric field over the entire aperture plane. The complete far field in two dimensions, obtained by combining equations (2.86) and (2.91), is asymptotically for $kr \rightarrow \infty$

$$\mathbf{E}(r,\theta) = \mathbf{u}_\theta\, E_\theta\,(r,\theta) + \mathbf{u}_y\, E_y\,(r,\theta)$$

$$= \sqrt{\frac{j\lambda}{r}}\, \exp\,(-jkr)\, [\mathbf{u}_\theta\, F_{\mathrm{TM}}\,(\sin\theta) + \mathbf{u}_y\, \cos\theta\, F_{\mathrm{TE}}\,(\sin\theta)] \tag{2.93}$$

and

$$\mathbf{H}\,(r,\theta) = Z^{-1}\, \mathbf{u}_r \times \mathbf{E}(r,\theta) \tag{2.94}$$

The situation is described pictorially in Fig. 2.21, which will be used to show that the fields represented by the two spectra $F_{\mathrm{TM}}\,(s)$ and $F_{\mathrm{TE}}\,(s)$ are not only complete but are also unique. The uniqueness theorem of Appendix B section B.2.9 applies to a source-free region such as the half-cylinder defined in Fig. 2.21 by the aperture plane and the circle of radius r. The theorem states

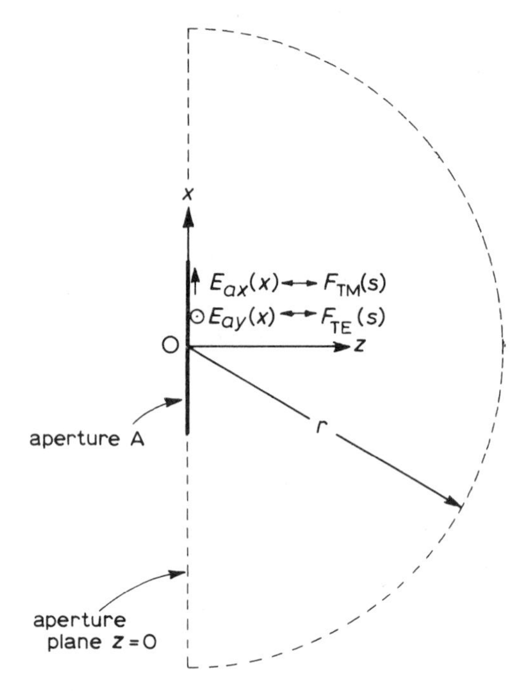

Fig. 2.21 – Complete representation of the two-dimensional field radiated from the aperture A into the half-space $z \geqslant 0$.

that if either the tangential electric or tangential magnetic field is specified uniquely at all points on the surface enclosing such a region, then the corresponding field throughout the entire region is itself unique. Over the aperture plane the tangential electric field has to be completely specified before the angular spectra $F_{TM}(s)$ and $F_{TE}(s)$ can be determined. For large r the field over the curved surface of the cylinder will have the far-field form of equations (2.93) and (2.94); the field components are everywhere tangential to the surface; as r tends to infinity the amplitudes of all field components tend to zero. Hence the two-dimensional field in the half-space $z \geqslant 0$ represented in terms of the two angular plane-wave spectra $F_{TM}(s)$ and $F_{TE}(s)$ by equations (2.84) and (2.87) is unique.

Exercise

The above uniqueness argument is incomplete. Examine the details of the uniqueness theorem itself, showing that when it is applied to fields of the form of equations (2.93) and (2.94) it is necessary to make the further assumption that the medium is very slightly lossy (that is, that k has a small negative imaginary part) which then completes the argument.

In this chapter we have treated the angular-spectrum representation of fields radiating in two dimensions in a fair amount of detail. There are two reasons for doing this. One is that, owing to the relative algebraic and geometrical simplicity of the two-dimensional development, the physical principles underlying the representation emerge with greater clarity. The other reason is that it often happens in three-dimensional radiation problems that the two-dimensional aperture field can be expressed as the product of two one-dimensional distributions (that is, the aperture field is separable). Then, as will be demonstrated in section 3.2.1 of the next chapter, the angular spectrum representing the three-dimensional radiated field is the product of the two angular spectra appropriate to the two one-dimensional aperture field distributions.

Exercise

Calculate the angular spectrum for the Gaussian aperture-field distribution

$$E_{ax}(x) = E_0 \exp\left(-\frac{x^2}{w_0^2}\right) \tag{2.95}$$

in which $x = \pm w_0$ are the points at which the bell-shaped curve reaches e^{-1} of its maximum value. Use the standard result (Dennery and Krzywicki (1967) p. 58)

$$\int_{-\infty}^{\infty} \exp(-\alpha x^2 \pm j\beta x)\,dx = \sqrt{\frac{\pi}{\alpha}} \exp\left(-\frac{\beta^2}{4\alpha}\right) \quad \text{for Re } \alpha > 0 \tag{2.96}$$

Exercise

Find the angular spectrum for the symmetrical triangular aperture-field distribution

$$E_{ax}(x) = E_0 \left(1 - \frac{2|x|}{b}\right) \quad \text{for } |x| \leqslant b/2$$
$$= 0 \text{ otherwise} \tag{2.97}$$

Exercise

Compare and contrast the far fields corresponding to the Gaussian, triangular, and rectangular aperture-field distributions. Pay particular attention to the on-axis field, the width of the main lobe, and the level of the sidelobes. In order to make a valid comparison, some form of normalization of the aperture fields will be required, such as that of equal radiated power or equal widths of an effective rectangular aperture distribution in each case.

Exercise

Establish the principle of holography by considering the intensity over the plane of the hologram of a single plane wave of the angular spectrum describing the scene of interest together with a coherent reference plane wave of much larger amplitude. Assume that the hologram, when developed, acquires an amplitude transfer characteristic proportional to the above intensity, so that when the developed hologram is illuminated by the reference wave alone the original scene is reconstructed.

CHAPTER 3

Three-dimensional fields

Having established the technique, in the previous chapter, of representing two-dimensional radiating fields by an angular spectrum of plane waves, the same technique will now be applied to three-dimensional fields.

3.1 ANGULAR SPECTRUM FOR LINEARLY POLARIZED APERTURE FIELDS

Consider the radiating aperture A of Fig. 3.1, defined as that region of the aperture plane $z = 0$ over which the tangential field is non-zero. If the tangential electric field in the aperture is wholly x-directed, then E_y is identically zero over the entire aperture plane and hence is zero everywhere in the half-space $z \geqslant 0$. The five remaining field components in $z \geqslant 0$ are in general non-zero.

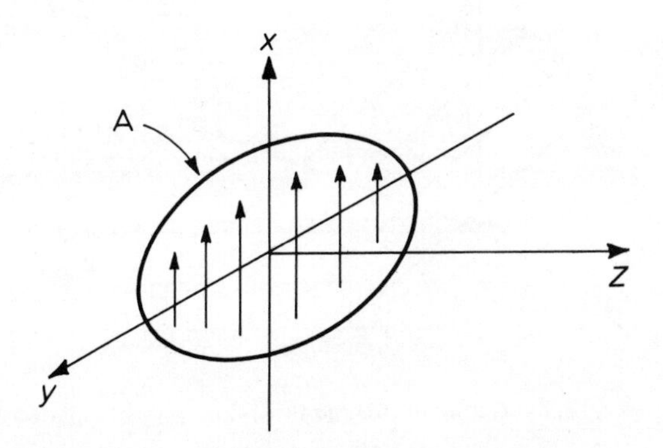

Fig. 3.1 – The aperture A : that part of the aperture plane over which the tangential electric field is non-zero.

We wish to construct an angular spectrum of plane waves, which will describe the field radiated from the aperture A into the half-space $z \geqslant 0$, based on a knowledge of the tangential component of the electric field in the aperture plane. Let this be

$$E_{ax}(x,y) = E_x(x,y,0) \quad , \tag{3.1}$$

and let the corresponding angular plane-wave spectrum function be $F_x(\alpha,\beta)$.

The direction **u** (See Fig. 3.2) of a typical elemental plane-wave component of the angular spectrum is conveniently specified by the direction cosines (α,β,γ), which are the components of **u** along the Cartesian axes. Thus

$$\mathbf{u} = \mathbf{u}_x\alpha + \mathbf{u}_y\beta + \mathbf{u}_z\gamma \tag{3.2}$$

and, since **u** is of the unit length,

$$\alpha^2 + \beta^2 + \gamma^2 = 1 \tag{3.3}$$

This means that if any two direction cosines are known the third follows automatically, which justifies specifying the angular spectrum $F_x(\alpha,\beta)$ in terms of the first two direction cosines.

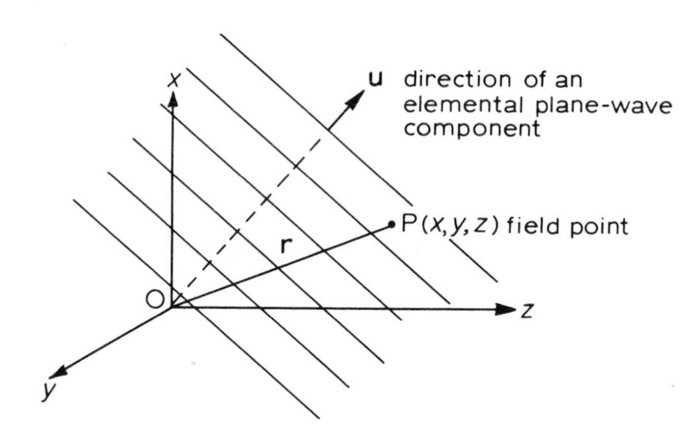

Fig. 3.2 – Geometry for the three-dimensional radiated field.

Exercise

An alternative method of specifying the direction **u** is in terms of polar angle θ and azimuth angle ϕ of Fig. 3.3. The polar angle θ is the angle between the direction of **u** and the z-axis. The azimuthal plane is the x-y plane, and the

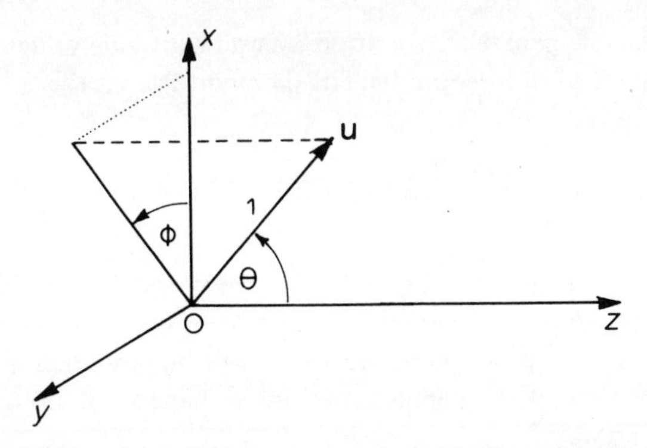

Fig. 3.3 – Direction **u** specified by spherical polar angle coordinates (θ,ϕ).

azimuth angle ϕ is taken as the angle between the directions of the projection of **u** on to the azimuthal plane and the x-axis. Show that the direction cosines are

$$\begin{aligned}\alpha &= \sin\theta\cos\phi\\ \beta &= \sin\theta\sin\phi\\ \gamma &= \cos\theta\end{aligned} \tag{3.4}$$

and confirm that $\alpha^2 + \beta^2 + \gamma^2 = 1$.

Now define the angular spectrum $F_x(\alpha,\beta)$ to be such that the x-component of the electric field of the elemental plane wave travelling in the direction (α,β) is $F_x(\alpha,\beta)\,\mathrm{d}\alpha\mathrm{d}\beta$. Then it follows from equation (2.14) that the elemental contribution to the x-component of the electric field at the point P in Fig. 3.2, distance **r** from O, is

$$\mathrm{d}E_x(\mathbf{r}) = F_x(\alpha,\beta)\,\mathrm{d}\alpha\mathrm{d}\beta\,\exp(-jk\mathbf{u}.\mathbf{r}) \tag{3.5}$$

In rectangular coordinates

$$\mathbf{r} = \mathbf{u}_x x + \mathbf{u}_y y + \mathbf{u}_z z$$

and, forming its scalar product with **u** in equation (3.2),

$$\mathbf{u}.\mathbf{r} = \alpha x + \beta y + \gamma z \tag{3.6}$$

so that

$$\mathrm{d}E_x(x,y,z) = F_x(\alpha,\beta)\,\mathrm{d}\alpha\mathrm{d}\beta\,\exp\{-jk(\alpha x + \beta y + \gamma z)\} \tag{3.7}$$

Integrating this elemental contribution over all allowable values of α and β yields the x-component of the electric field at the point $P(x,y,z)$ as

$$E_x(x,y,z) = \int_{-\infty}^{\infty}\int_{-\infty}^{\infty} F_x\,(\alpha,\beta)\exp\{-jk\,(\alpha x + \beta y + \gamma z)\}\,\mathrm{d}\alpha\mathrm{d}\beta\ . \tag{3.8}$$

The significance of including all real values of α and β is that the angular plane-wave spectrum $F_x(\alpha,\beta)$ thereby incorporates non-propagating (evanescent) waves in addition to propagating waves, as will be described in detail in section 3.3. The remaining field components are obtained, all in terms of $F_x(\alpha,\beta)$, in the following way.

Returning to the elemental plane wave of equation (3.7), the electric field has zero y-component, by assumption, and hence (using the plane-wave condition of equation (2.15), that $\mathbf{u.E} = 0$) the Cartesian components of the electric field are in the ratio

$$\mathrm{d}E_x : \mathrm{d}E_y : \mathrm{d}E_z = 1 : 0 : -\frac{\alpha}{\gamma}\ . \tag{3.9}$$

Having obtained the electric field, the magnetic-field components of the elemental plane wave must bear the following ratio to the x-component of the electric field (by applying the plane-wave condition of equation (2.16), that $Z\mathbf{H} = \mathbf{u} \times \mathbf{E}$)

$$\mathrm{d}E_x : \mathrm{d}H_x : \mathrm{d}H_y : \mathrm{d}H_z = 1 : -\frac{\alpha\beta}{\gamma Z} : \frac{1-\beta^2}{\gamma Z} : \frac{-\beta}{Z} \tag{3.10}$$

where Z is the plane-wave impedance of the medium.

Exercise

Check the validity of the ratios (3.9) and (3.10).

Hence the electric and magnetic fields at the point (x,y,z) in the half-space $z \geqslant 0$, due to the aperture field $E_{ax}(x,y)$, are

$$\mathbf{E}(x,y,z) = \int_{-\infty}^{\infty}\int_{-\infty}^{\infty} [\mathbf{u}_x - \mathbf{u}_z \frac{\alpha}{\gamma}]\ F_x(\alpha,\beta)\ \exp\{-jk\,(\alpha x + \beta y + \gamma z)\}\,\mathrm{d}\alpha\mathrm{d}\beta \tag{3.11}$$

and

$$\mathbf{H}(x,y,z) = \frac{1}{Z}\int_{-\infty}^{\infty}\int_{-\infty}^{\infty} [-\mathbf{u}_x \frac{\alpha\beta}{\gamma} + \mathbf{u}_y \frac{1-\beta^2}{\gamma} - \mathbf{u}_z\,\beta]\ F_x(\alpha,\beta)\ \exp\{-jk\,(\alpha x + \beta y + \gamma z)\}\,\mathrm{d}\alpha\mathrm{d}\beta\ . \tag{3.12}$$

Note that all five field components have been expressed in terms of the single angular-spectrum function $F_x(\alpha,\beta)$. The sixth component, E_y, is zero everywhere in $z \geqslant 0$.

Exercise

Show that the electromagnetic field of equations (3.11) and (3.12) satisfies Maxwell's equations.

Now consider the orthogonal linearly polarized aperture field. That is, the tangential electric field over the aperture plane is wholly y-directed, and E_x is zero everywhere in $z \geqslant 0$. Let the aperture field in this case be

$$E_{ay}(x,y) = E_y(x,y,0) \tag{3.13}$$

and let the corresponding angular plane-wave spectrum function for this y-polarized aperture field be $F_y(\alpha,\beta)$.

By means of a construction similar to that described above for the x-polarized aperture field, the electric and magnetic fields at the point (x,y,z) in the half-space $z \geqslant 0$ due to the aperture field $E_{ay}(x,y)$ are

$$\mathbf{E}(x,y,z) = \int_{-\infty}^{\infty}\int_{-\infty}^{\infty} [\mathbf{u}_y - \mathbf{u}_z \frac{\beta}{\gamma}]\; F_y(\alpha,\beta) \exp(-jk\mathbf{u}.\mathbf{r})\, \mathrm{d}\alpha \mathrm{d}\beta \tag{3.14}$$

and

$$\mathbf{H}(x,y,z) = \frac{1}{Z}\int_{-\infty}^{\infty}\int_{-\infty}^{\infty} [-\mathbf{u}_x \frac{1-\alpha^2}{\gamma} + \mathbf{u}_y \frac{\alpha\beta}{\gamma} + \mathbf{u}_z\, \alpha]\; F_y(\alpha,\beta) \exp(-jk\mathbf{u}.\mathbf{r})\, \mathrm{d}\alpha \mathrm{d}\beta \tag{3.15}$$

in which the exponential phase factor has been written in vector form for brevity. Note that we again have five non-zero field components expressed in terms of the single angular spectrum function $F_y(\alpha,\beta)$.

Exercise

By means of a suitable coordinate transformation derive equations (3.14) and (3.15) from equations (3.11) and (3.12).

Exercise

Use the principle of duality to show that the field radiated into the half-space $z \geqslant 0$ can be represented by two angular spectrum functions, based on two orthogonal, linearly polarized components of the aperture *magnetic* field.

Since any aperture field $\mathbf{E}_a(x,y)$, which is the tangential component of the electric field over the aperture plane $z \geqslant 0$, can be written as the sum of two orthogonal, linearly polarized aperture fields, namely,

$$\mathbf{E}_a(x,y) = \mathbf{u}_x\, E_{ax}(x,y) + \mathbf{u}_y\, E_{ay}(x,y) \tag{3.16}$$

it is clear that we have established the following general result. The radiation into the half-space bounded by a plane over which the tangential electric (or magnetic) field is known can be represented by two angular spectrum functions. By superposition, the complete electric field is given by the sum of equations (3.11) and (3.14), and the complete magnetic field by the sum of equations (3.12) and (3.15). However, in order to reduce the amount of algebra to be carried along we will usually deal with only one of the orthogonally polarized aperture fields, the implication being that the other polarization can be dealt with similarly. Indeed it will often be sufficient to investigate a single component, such as E_x, of one polarization of the aperture field, since all field components are similar in form.

The method of deriving the angular plane-wave spectra from the aperture field distribution is given next.

3.2 APERTURE FIELD: FOURIER TRANSFORM OF THE ANGULAR SPECTRUM

For the x-polarized aperture field, depicted in Fig. 3.1 and defined by equation (3.1), substitution of equation (3.8) yields

$$E_{ax}(x,y) = \int_{-\infty}^{\infty}\int_{-\infty}^{\infty} F_x(\alpha,\beta)\exp\{-jk(\alpha x + \beta y)\}\, d\alpha\, d\beta \tag{3.17}$$

Thus the aperture field $E_{ax}(x,y)$ is the two-dimensional Fourier transform of the angular plane-wave spectrum $F_x(\alpha,\beta)$, or symbolically

$$F_x(\alpha,\beta) \longleftrightarrow E_{ax}(x,y) \tag{3.18}$$

Inverting equation (3.17),

$$F_x(\alpha,\beta) = \frac{1}{\lambda^2}\int_{-\infty}^{\infty}\int_{-\infty}^{\infty} E_{ax}(x,y)\exp\{+jk(\alpha x + \beta y)\}\, dx\, dy \tag{3.19}$$

establishes how the angular spectrum may be derived once the aperture field is known.

The y-polarized aperture field $E_{ay}(x,y)$ is, by the same argument, the two-dimensional Fourier transform of its corresponding angular plane-wave spectrum $F_y(\alpha,\beta)$, that is,

$$F_y(\alpha,\beta) \longleftrightarrow E_{ay}(x,y) \tag{3.20}$$

3.2.1 Separable Aperture Fields

If the aperture field, such as $E_{ax}(x,y)$, can be expressed as the product of two functions, one of which depends on x alone and the other on y alone, the aperture field is said to be separable, and considerable simplification follows. Let

$$E_{ax}(x,y) = f_1(x) f_2(y) \quad , \tag{3.21}$$

then the corresponding angular spectrum is, from equation (3.19),

$$F_x(\alpha,\beta) = \frac{1}{\lambda^2} \int_{-\infty}^{\infty} f_1(x) \exp(jk\alpha x)\, dx \int_{-\infty}^{\infty} f_2(y) \exp(jk\beta y)\, dy \tag{3.22}$$

or

$$F_x(\alpha,\beta) = F_1(\alpha) F_2(\beta) \tag{3.23}$$

Thus, when the aperture field is separable the angular spectrum is also separable, and the factors $F_1(\alpha)$ and $F_2(\beta)$ into which it separates are simply the one-dimensional spectra corresponding to the one-dimensional factors $f_1(x)$ and $f_2(y)$ of the aperture field, namely,

$$F_1(\alpha) = \frac{1}{\lambda} \int_{-\infty}^{\infty} f_1(x) \exp(jk\alpha x)\, dx \tag{3.24}$$

and

$$F_2(\beta) = \frac{1}{\lambda} \int_{-\infty}^{\infty} f_2(y) \exp(jk\beta y)\, dy \tag{3.25}$$

which have exactly the same form as equation (2.43).

Example

Consider the radiation from a rectangular aperture (see Fig. 3.4) cut from a perfectly conducting sheet coincident with the plane $z = 0$. The aperture field will be zero outside the rectangular aperture. Suppose that within the aperture

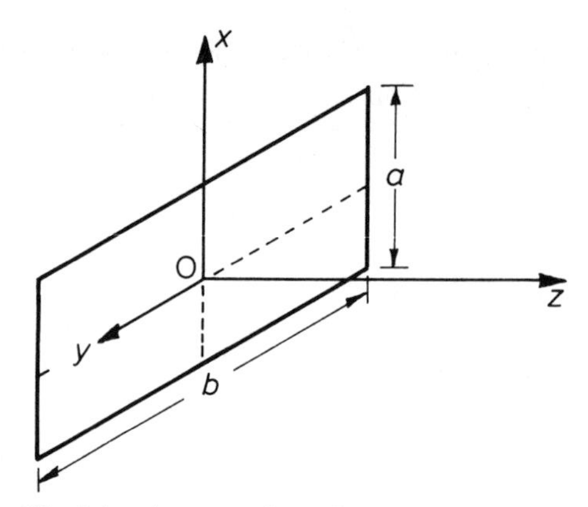

Fig. 3.4 – A rectangular radiating aperture.

the tangential electric field is wholly x-directed, uniform in phase, with an amplitude that is uniform in the x-direction but has the form of a half-cosine in the y-direction. We may therefore write the aperture field as

$$E_{ax}(x,y) = E_0 \cos\left(\frac{\pi y}{b}\right) \quad \text{for} \quad \begin{array}{l} |x| \leqslant a/2 \\ |y| \leqslant b/2 \end{array}$$
$$= 0 \text{ otherwise,} \tag{3.26}$$

which is clearly separable. Note that equation (3.26) describes the field over the cross-section of a rectangular metal waveguide which supports the fundamental TE_{01} mode (see Appendix A.10).

For the separable aperture field of equation (3.26) choose the factors as

$$f_1(x) = E_0 \text{ for } |x| \leqslant a/2$$
$$= 0 \quad \text{otherwise,} \tag{3.27}$$

and

$$f_2(y) = \cos\left(\frac{\pi y}{b}\right) \text{ for } |y| \leqslant b/2$$
$$= 0 \quad \text{otherwise.} \tag{3.28}$$

The corresponding one-dimensional spectra, obtained by substituting equations (3.27) and (3.28) into equations (3.24) and (3.25), are

$$F_1(\alpha) = \frac{E_0}{\lambda} \int_{-a/2}^{a/2} \exp(jk\alpha x)\, dx = E_0 \frac{a}{\lambda} \operatorname{sinc}\left(\frac{a\alpha}{\lambda}\right) \tag{3.29}$$

(cf. equations (2.49) and (2.50) and

$$F_2(\beta) = \frac{1}{2\lambda}\int_{-b/2}^{b/2}\left[\exp\left(j\frac{\pi y}{b}\right) + \exp\left(-j\frac{\pi y}{b}\right)\right]\exp(jk\beta y)\,dy$$

$$= \frac{b}{2\lambda}\left\{\text{sinc}\left[\frac{b}{\lambda}\left(\beta + \frac{\lambda}{2b}\right)\right] + \text{sinc}\left[\frac{b}{\lambda}\left(\beta - \frac{\lambda}{2b}\right)\right]\right\}$$

$$= \frac{\pi b}{2\lambda}\frac{\cos\left(\dfrac{\pi b\beta}{\lambda}\right)}{\left(\dfrac{\pi}{2}\right)^2 - \left(\dfrac{\pi b\beta}{\lambda}\right)^2}\,. \tag{3.30}$$

Hence the complete angular spectrum for the aperture field of equation (3.26) is the product of equations (3.29) and (3.30), namely,

$$F_x(\alpha,\beta) = E_0\frac{\pi ab}{2\lambda^2}\frac{\sin\left(\dfrac{\pi a\alpha}{\lambda}\right)}{\dfrac{\pi a\alpha}{\lambda}}\frac{\cos\left(\dfrac{\pi b\beta}{\lambda}\right)}{\left(\dfrac{\pi}{2}\right)^2 - \left(\dfrac{\pi b\beta}{\lambda}\right)^2}\,. \tag{3.31}$$

Based on our experience with two-dimensional fields, and anticipating its demonstration in section 3.4, we can identify $F_x(\alpha,\beta)$ as the far-field pattern of the radiation. Fig. (3.5) gives a comparison of the patterns in the two principal planes: one in the x-z plane ($\beta = 0$) and the other in the y-z plane ($\alpha = 0$).

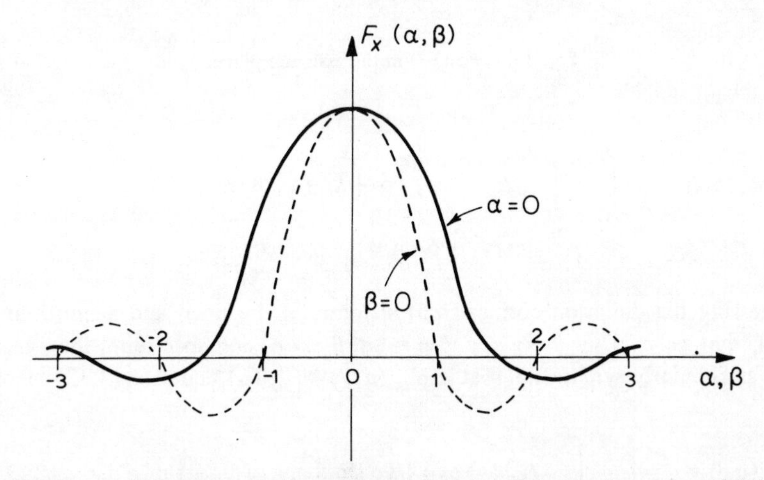

Fig. 3.5 – Plot of the angular spectrum function $F_x(\alpha,\beta)$ of equation (3.31) with $a = b = \lambda$.

Note how, for the same aperture width, the half-cosine aperture field in the y-direction produces a broader pattern with less pronounced sidelobes than does the uniform aperture field in the x-direction.

Exercise

Show that the three-dimensional field representation reduces to the two-dimensional form considered previously when the aperture field is taken to be uniform and of infinite extent in, say, the y-direction.

3.2.2 Circularly Symmetric Aperture Fields

If the aperture field is simply a function of the radial distance from the origin, that is, it is circularly symmetric, the three-dimensional radiated field can be represented in terms of a single one-dimensional angular spectrum, as the following argument shows. Fig. (3.6) is a view of the aperture plane seen from behind, looking out into the half-space $z \geqslant 0$. A point Q in the aperture plane can be represented by the circular polar coordinates (ρ,ψ). The angular spectrum of equation (3.19) then becomes

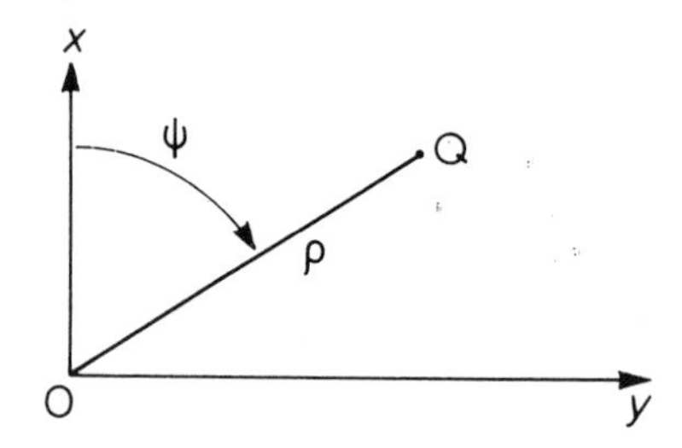

Fig. 3.6 – Point Q in the aperture plane.

$$F_x(\alpha,\beta) = \frac{1}{\lambda^2}\int_0^{\infty}\int_0^{2\pi} E_{ax}(\rho,\psi)\exp\{jk\rho\,(\alpha\cos\psi + \beta\sin\psi)\}\,\rho d\rho\, d\psi \ . \qquad (3.32)$$

Expressing the direction cosines (α,β) in terms of the polar and azimuth angles (θ,ϕ), that is, $\alpha = \sin\theta\cos\phi$ and $\beta = \sin\theta\sin\phi$, and specifying the aperture field as circularly symmetric, that is, $E_{ax}(\rho,\psi) = E_{ax}(\rho)$, equation (3.32) becomes

$$F_x(\alpha,\beta) = \frac{1}{\lambda^2}\int_0^{\infty}\int_0^{2\pi} E_{ax}(\rho)\exp\{jk\rho\sin\theta\cos(\psi-\phi)\}\,\rho d\rho\, d\psi \ . \qquad (3.33)$$

Then applying the standard result (see p. 193) that

$$\int_0^{2\pi} \exp(jq\cos\theta)\,\mathrm{d}\theta = 2\pi\, J_0(q) \quad , \tag{3.34}$$

in which $J_0\ (\ )$ is the Bessel function of the first kind and zero order, equation (3.33) becomes

$$F_x(s) = \frac{2\pi}{\lambda^2}\int_0^{\infty} E_{ax}(\rho)\, J_0\,(k\rho s)\,\rho\mathrm{d}\rho \quad , \tag{3.35}$$

where the angular spectrum is a function of the single variable $s = \sin\theta$. Hence one can say that the angular spectrum for a circularly symmetric apeture-field distribution is itself circularly symmetric (that is, independent of the azimuth angle ϕ).

Example

Consider the uniformly illuminated circular aperture of radius a, for which the aperture field

$$\begin{aligned} E_{ax}(\rho) &= E_0 \quad \text{for } \rho \leqslant a \\ &= 0 \quad \text{otherwise,} \end{aligned} \tag{3.36}$$

is circularly symmetric. The angular spectrum describing the radiation from this aperture is therefore

$$F_x(s) = \frac{2\pi E_0}{\lambda^2}\int_0^{a} J_0(k\rho s)\,\rho\mathrm{d}\rho \tag{3.37}$$

Using another standard result for Bessel functions, that

$$\int_0^{a} x\, J_0(x)\,\mathrm{d}x = \Big[x\, J_1(x)\Big]_0^{a} = a\, J_1(a) \quad , \tag{3.38}$$

in which $J_1\ (\ )$ is the Bessel function of first kind and first order, the angular spectrum for the uniformly illuminated circular aperture is

$$F_x(s) = E_0\,\frac{\pi a^2}{\lambda^2}\,\frac{2J_1(kas)}{kas} \tag{3.39}$$

The final quotient in this equation is similar in form to the sinc function, to which it is compared in Fig. 3.7.

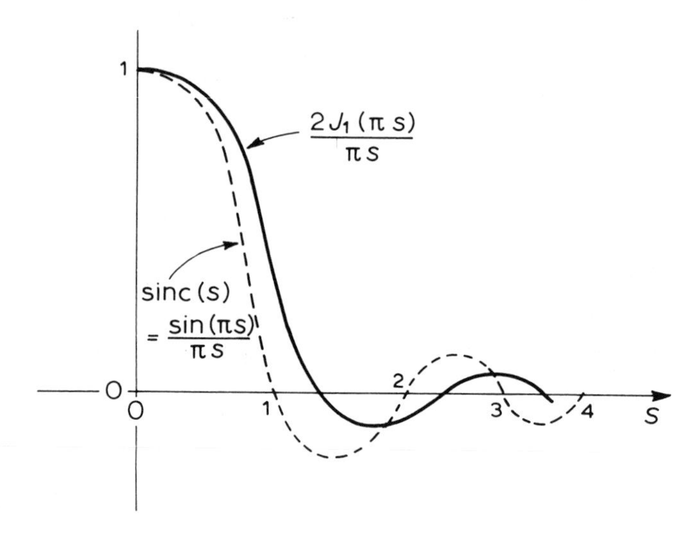

Fig. 3.7 – Comparison of the patterns, as a function of $s = \sin\theta$, of the uniformly illuminated slit (width $a = \lambda$) and uniformly illuminated circular aperture (diameter $2a = \lambda$).

Exercise

Find the angular spectrum for the two-dimensional Gaussian aperture field distribution

$$E_{ax}(x,y) = E_0 \exp\left(-\frac{x^2 + y^2}{w_0^2}\right) \tag{3.40}$$

in which w_0 is a constant. (Note that equation (3.40) describes the transverse field of the output of a laser in its fundamental mode, and is also used as an idealized model of a tapered antenna aperture field.)

3.3 EVANESCENT WAVES : AND ALTERNATIVE REPRESENTATIONS

The representation of field components such as that of equation (3.8):

$$E_x(x,y,z) = \int_{-\infty}^{\infty}\int_{-\infty}^{\infty} F_x(\alpha,\beta) \exp\{-jk(\alpha x + \beta y + \gamma z)\}\, d\alpha d\beta \tag{3.41}$$

as an integral over all real values of α and β (the direction cosines) means that the representation includes evanescent waves. Consider the elemental plane-wave contribution to the field in equation (3.41).

$$dE_x(x,y,z) = F_x(\alpha,\beta)\ d\alpha d\beta \exp\{-jk(\alpha x + \beta y + \gamma z)\} \tag{3.42}$$

This is a plane wave travelling in the half-space $z \geqslant 0$. If $\alpha^2 + \beta^2 \leqslant 1$, the third direction cosine γ is real, and must be chosen to be positive so that the plane wave travels into the region $z \geqslant 0$. If $\alpha^2 + \beta^2 > 1$, then γ is purely imaginary, and must be chosen to be negative in order that the fields remain finite as $z \to +\infty$. When γ is negative imaginary the plane wave no longer has a component of propagation in the z direction: indeed the wave merely attenuates in that direction. The wave is therefore said to be evanescent, as described already in connection with two-dimensional fields in section 2.3. The wave does propagate, but always in a direction parallel to the aperture plane. The energy associated with these waves is therefore tied to the aperture plane and can be thought of as being stored there. For this reason the field due to the evanescent waves is often referred to as the **reactive field**. Another feature of evanescent waves is that the amplitude is not constant across their wavefronts, defined as the planes of constant phase. As a result they are also referred to as **inhomogeneous plane waves.**

The circle $\alpha^2 + \beta^2 = 1$ (see Fig. 3.8) is the boundary between the two domains in the α-β plane, the interior of the circle corresponding to propagating waves while the exterior corresponds to evanescent waves.

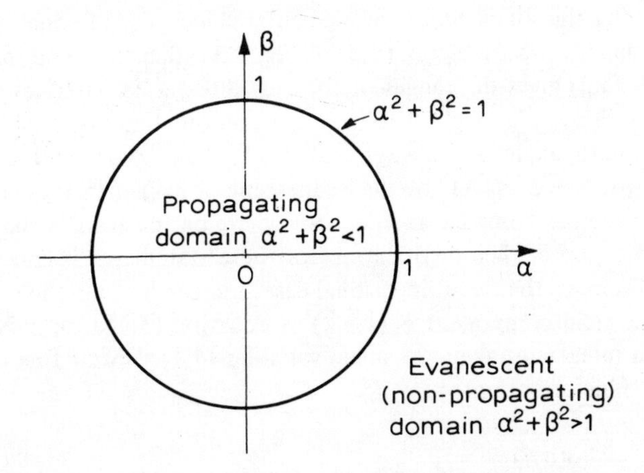

Fig. 3.8 – Domains in the α-β plane of propagating and evanescent waves.

3.3.1 Wavenumber-component representation

A popular variant of the direction-cosine representation of equation (3.41) is obtained by writing

$$k_x = k\alpha, \quad k_y = k\beta, \quad k_z = k\gamma \tag{3.45}$$

which are the Cartesian components of the vector wavenumber **k** (see equation (2.18)), and for which

$$k_x^2 + k_y^2 + k_z^2 = k^2 \quad . \tag{3.44}$$

Then, in terms of the first two components k_x and k_y,

$$E_x(x,y,z) = k^{-2} \int_{-\infty}^{\infty} \int_{-\infty}^{\infty} F_x(k_x,k_y) \exp\{-j(k_x x + k_y y + k_z z)\} \, dk_x \, dk_y \tag{3.45}$$

$$= k^{-2} \int_{-\infty}^{\infty} \int_{-\infty}^{\infty} F_x(k_x,k_y) \exp\{-j\mathbf{k}.\mathbf{r}\} \, dk_x \, dk_y \quad . \tag{3.46}$$

The domain of propagating waves in the $k_x - k_y$ plane is the interior of the circle $k_x^2 + k_y^2 = k^2$, of radius k. Outside this circle, $k_x^2 + k_y^2 > k^2$, and the waves are evanescent.

3.3.2 Spherical-polar Angle Representation

Specifying the direction of the elemental plane wave of equation (3.42) in terms of the spherical-polar angles (θ,ϕ) of Fig. 3.3 rather than in terms of the direction cosines (α,β) gives the ranges of the new integration variables θ and ϕ shown in Fig. 3.9. The polar angle θ follows the contour Γ in the complex θ plane, while the azimuth angle ϕ is always real. Both θ and ϕ are real in the propagating domain $(\alpha^2 + \beta^2 \leqslant 1)$. In the evanescent domain $(\alpha^2 + \beta^2 > 1)$ ϕ is still real, but θ becomes complex and lies somewhere on the arms of the Γ-contour which extend to $\pm j\,\infty$. The justification for these statements is completely analogous to that used in the two-dimensional case, and can be found in section 2.2.

The field component $E_x(x,y,z)$ of equation (3.41) can now be transformed into an integration over the angle variables (θ,ϕ). Noting first that the Jacobean of the transformation

$$\frac{\partial(\alpha,\beta)}{\partial(\theta,\phi)} = \sin\theta \cos\theta \tag{3.47}$$

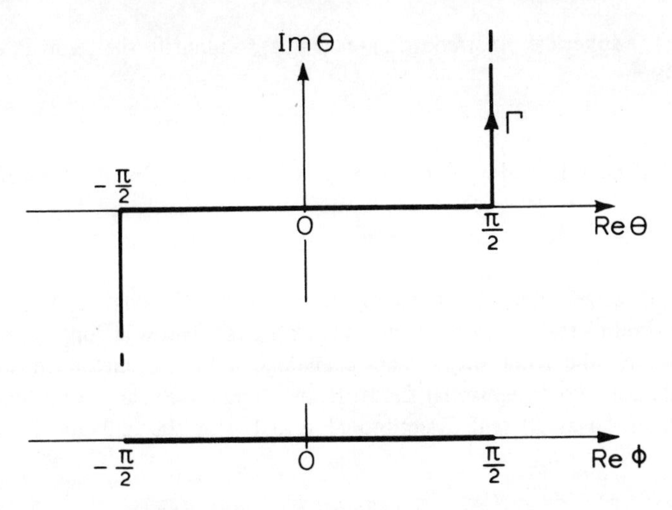

Fig. 3.9 – Ranges of the integration variables (θ,ϕ).

it follows that

$$E_x(x,y,z) = \int_\Gamma \mathrm{d}\theta \int_{-\pi/2}^{\pi/2} \mathrm{d}\phi \, F_x(\sin\theta\cos\phi, \sin\theta\sin\phi)$$

$$\sin\theta\cos\theta \exp\{-jk(x\sin\theta\cos\phi + y\sin\theta\sin\phi + z\cos\theta)\} \,. \quad (3.48)$$

3.4 FAR FIELD : APPROXIMATED BY THE ANGULAR SPECTRUM

So far we have spoken only of the *representation* of the field at some point P in the half-space $z \geqslant 0$ in terms of the angular plane-wave spectra $F_x(\alpha,\beta)$ and $F_y(\alpha,\beta)$. Now we turn to its *solution*, which involves the evaluation of double integrals of the type

$$E_x(x,y,z) = \int_{-\infty}^{\infty}\int_{-\infty}^{\infty} F_x(\alpha',\beta') \exp\{-jk(\alpha'x + \beta'y + \gamma'z)\} \,\mathrm{d}\alpha'\mathrm{d}\beta' \quad (3.49)$$

In this integral the direction cosines α', β', which are the integration variables, have been written with primes in order to distinguish them from α,β which will now be used to represent the fixed direction to the field point. Thus

$$x = r\alpha, \quad y = r\beta, \quad z = r\gamma \quad (3.50)$$

and using the spherical polar coordinates (r,θ,ϕ) to identify the point P, equation (3.49) becomes

$$E_x(r,\theta,\phi) = \int_{-\infty}^{\infty}\int_{-\infty}^{\infty} F_x(\alpha',\beta') \exp\{-jkr\,(\alpha'\alpha + \beta'\beta + \gamma'\gamma)\}\, d\alpha' d\beta' \tag{3.51}$$

If kr is very large, that is, $r \gg \lambda$, this integral can be evaluated approximately by the method of stationary phase and yields the far-field solution.

Physically, the situation is exactly analogous to that already described in section 2.5 for two-dimensional fields. Evanescent waves can be neglected, and as α', β' range over all real directions (Fig. 3.10) in the half-space $z \geqslant 0$ the

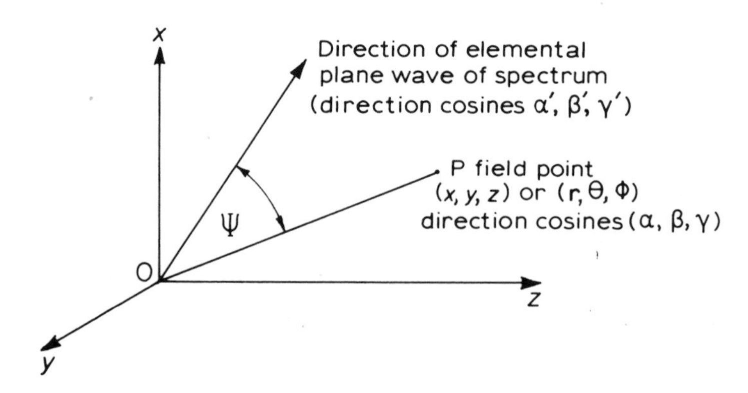

Fig. 3.10 – Geometry for the stationary-phase evaluation of the field at P.

phase of the integrand of equation (3.51), which can be written in terms of the angle Ψ between the two directions (α',β') and (α,β) as

$$kr\,(\alpha'\alpha + \beta'\beta + \gamma'\gamma) = kr \cos \Psi, \tag{3.52}$$

changes very rapidly, except when $\cos \Psi$ is stationary. This condition of the phase being stationary clearly occurs when $\Psi = 0$ and $\alpha' = \alpha$, $\beta' = \beta$, $\gamma' = \gamma$, which is when the direction of the elemental plane wave coincides with the direction to the field point. If the angular spectrum function $F_x(\alpha',\beta')$ is continuous and bounded, which must be so if the aperture A is finite, then kr can always be chosen large enough that all contributions from the integrand effectively cancel each other, except in the direction for which the phase is stationary. Hence equation (3.51) can be written as the approximate proportionality

$$E_x(r,\theta,\phi) \sim F_x(\sin\theta\cos\phi, \sin\theta\sin\phi) \exp(-jkr) \tag{3.53}$$

which shows that the far field is approximately proportional to the angular plane-wave spectrum. The precise form of this relationship will now be developed with the aid of the following algorithm.

3.4.1 Stationary-phase Evaluation of Double Integrals

Consider the double integral

$$I = \int_D f(x,y) \exp\{jKg(x,y)\} \mathrm{d}x\,\mathrm{d}y \tag{3.54}$$

in which the functions $f(x,y)$ and $g(x,y)$ are real and continuous over the domain of integration D, and K is a large positive real number. Assume that $g(x,y)$ is stationary at the single point (x_0,y_0) within D, so that

$$g_x(x_0,y_0) = \left(\frac{\partial g}{\partial x}\right)_{\substack{x=x_0\\y=y_0}} = 0 \tag{3.55}$$

and

$$g_y(x_0,y_0) = \left(\frac{\partial g}{\partial y}\right)_{\substack{x=x_0\\y=y_0}} = 0 \ .$$

where the subscripts are used here to denote partial differentiation. Expanding $g(x,y)$ in a Taylor series about the stationary point (x_0,y_0),

$$g(x,y) = g(x_0,y_0) + \tfrac{1}{2}[\xi^2 g_{xx}(x_0,y_0) + 2\xi\eta\, g_{xy}(x_0,y_0) + \eta^2 g_{yy}(x_0,y_0)] + \ldots . \tag{3.56}$$

in which $\xi = x - x_0$, $\eta = y - y_0$, and $g_{xy} = \dfrac{\partial^2 g}{\partial x \partial y}$, for example.

Then according to the physical arguments given above, for large K the only significant contribution to the integral I comes from the neighbourhood of the stationary point. Hence I is given asymptotically for $K \to \infty$ by

$$I \simeq f(x_0,y_0) \exp\{jKg(x_0,y_0)\} \int_{-\infty}^{\infty}\int_{-\infty}^{\infty} \exp\{j\tfrac{1}{2}K\,[\xi^2 g_{xx} + 2\xi\eta\, g_{xy} + \eta^2 g_{yy}]\}\, \mathrm{d}\xi\,\mathrm{d}\eta \tag{3.57}$$

where it is understood that the partial derivatives are evaluated at the stationary point. By completing the square in the integrand and using the standard integral

$$\int_{-\infty}^{\infty} \exp(-ax^2 \pm jbx)\,dx = \sqrt{\frac{\pi}{a}} \exp\left(-\frac{b^2}{4a}\right) \tag{3.58}$$

which is valid for Re $a > 0$, equation (3.57) becomes

$$I \simeq \frac{\pm j2\pi}{K\,(g_{xx}\,g_{yy} - g_{xy}^2)^{1/2}}\, f(x_0,y_0) \exp\{jKg(x_0,y_0) \tag{3.59}$$

where the positive sign is taken, unless both $g_{xx} < 0$ and $g_{xx}\,g_{yy} > g_{xy}^2$.

Applying the above stationary-phase algorithm to the double integral of equation (3.51) the x-component of the electric field has the asymptotic value, as $kr \rightarrow \infty$, of

$$E_x(r,\theta,\phi) \simeq j\frac{2\pi}{kr} \exp(-jkr) \cos\theta\, F_x(\sin\theta\, \cos\phi, \sin\theta\, \sin\phi) \tag{3.60}$$

which is clearly a spherically-expanding wave whose amplitude in a particular direction depends directly on the value of the angular plane-wave spectrum function in that direction.

The remaining Cartesian components of the field radiated from an x-polarized aperture-field distribution, whose angular spectrum is $F_x\,(\alpha,\beta)$, can be obtained from equations (3.11) and (3.12) by a similar application of the stationary-phase principle. This gives the vector electric field

$$\mathbf{E}(r,\theta,\phi) = j\,\frac{2\pi}{kr} \exp(-jkr)\, F_x(\alpha,\beta)\, [\mathbf{u}_x\gamma - \mathbf{u}_z\alpha] \tag{3.61}$$

and the vector magnetic field as

$$\mathbf{H}\,(r,\theta,\phi) = j\,\frac{2\pi}{Zkr} \exp(-jkr)\, F_x(\alpha,\beta)$$

$$[-\mathbf{u}_x\,\alpha\beta + \mathbf{u}_y\,(1-\beta^2) - \mathbf{u}_z\,\beta\gamma] \tag{3.62}$$

both asymptotically as $kr \rightarrow \infty$. The angle factors have been condensed by writing them in terms of the direction cosines (α,β,γ) which refer here to the direction of the field point (r,θ,ϕ).

Exercise

If the Cartesian components (A_x, A_y, A_z) of a vector **A** are known, show that the spherical-polar components of **A** are given by

$$\begin{aligned} A_r &= A_x \sin\theta \cos\phi + A_y \sin\theta \sin\phi + A_z \cos\theta \\ A_\theta &= A_x \cos\theta \cos\phi + A_y \cos\theta \sin\phi - A_z \sin\theta \\ A_\phi &= -A_x \sin\phi + A_y \cos\phi \ . \end{aligned} \tag{3.63}$$

Show also the converse, that

$$\begin{aligned} A_x &= A_r \sin\theta \cos\phi + A_\theta \cos\theta \cos\phi - A_\phi \sin\phi \\ A_y &= A_r \sin\theta \sin\phi + A_\theta \cos\theta \sin\phi + A_\phi \cos\phi \\ A_z &= A_r \cos\theta - A_\theta \sin\theta \ . \end{aligned} \tag{3.64}$$

Converting the far field of equations (3.61) and (3.62) from Cartesian components into spherical-polar components, which are a more natural choice for a spherical wave, application of equation (3.63) yields

$$\mathbf{E}(r,\theta,\phi) = j\,\frac{2\pi}{kr}\exp(-jkr)\,F_x(\alpha,\beta)\,[\mathbf{u}_\theta \cos\phi - \mathbf{u}_\phi \cos\theta \sin\phi] \tag{3.65}$$

and

$$\mathbf{H}(r,\theta,\phi) = j\,\frac{2\pi}{Zkr}\exp(-jkr)\,F_x(\alpha,\beta)\,[\mathbf{u}_\theta \cos\theta \sin\phi - \mathbf{u}_\phi \cos\phi] \tag{3.66}$$

It is clear from equations (3.65) and (3.66) that the far field satisfies the two conditions

$$\mathbf{u}_r \,.\, \mathbf{E} = 0 \tag{3.67}$$

and

$$\mathbf{H} = \frac{1}{Z}\,\mathbf{u}_r \times \mathbf{E} \tag{3.68}$$

which show that the electromagnetic field in the region of a far field point has the *local* character of a plane wave cf. equations (2.15) and (2.16).

The power flow in the far field, given by the Poynting vector $\mathbf{S} = \frac{1}{2}\,\mathrm{Re}\,\mathbf{E} \times \mathbf{H}^*$, has a radial component only:

$$S_r(r,\theta,\phi) = \frac{\lambda^2}{2Zr^2}\,(1 - \beta^2)\,|F_x(\alpha,\beta)|^2 \tag{3.69}$$

Note that for angular spectra narrowly centred around the z-axis, the power flow is just proportional to the square modulus of the angular plane wave spectrum, and for this reason $|F_x(\alpha,\beta)|^2$ is referred to as the **angular power spectrum.**

3.4.2 Far Field for y-polarized Aperture Fields

If the electric field tangential to the aperture plane is wholly y-directed, the angular spectrum of the radiation $F_y(\alpha,\beta)$ is then the double Fourier transform of the aperture field $E_{ay}(x,y)$, and the far field is (asymptotically for $kr \to \infty$)

$$\mathbf{E}(r,\theta,\phi) = j\,\frac{2\pi}{kr}\,\exp(-jkr)\,F_y(\alpha,\beta)[\mathbf{u}_\theta \sin\phi + \mathbf{u}_\phi \cos\theta\cos\phi] \tag{3.70}$$

and

$$\mathbf{H}(r,\theta,\phi) = j\,\frac{2\pi}{Zkr}\,\exp(-jkr)\,F_y(\alpha,\beta)[-\mathbf{u}_\theta \cos\theta\cos\phi + \mathbf{u}_\phi \sin\phi]$$

$$= Z^{-1}\,\mathbf{u}_r \times \mathbf{E}\,(r,\theta,\phi)\quad . \tag{3.71}$$

The radial power flow in this case is

$$S_r(r,\theta,\phi) = \frac{\lambda^2}{2Zr^2}\,(1-\alpha^2)|F_y(\alpha,\beta)|^2 \quad . \tag{3.72}$$

Exercise

Start with the general specification of the field radiated by the y-polarized aperture field of equations (3.14) and (3.15) and derive the far field of equations (3.70) and (3.71).

Exercise

If the aperture field is circularly polarized, find out in what directions the far field is circularly polarized.

Exercise

Apply the uniqueness argument, given in section 2.6 for two-dimensional fields, to three-dimensional fields specified in terms of the two angular plane-wave spectra $F_x(\alpha,\beta)$ and $F_y(\alpha,\beta)$.

3.4.3 Representation for an arbitrary waveform

In order to broaden the picture it is worth a remark in passing (it will play no further part in our discussions) that if the aperture field is some arbitrary function of time, rather than the single-frequency, monochromatic, waveform assumed previously, say of the form $E_{ax}(x,y,t)$, then the field in the half-space $z \geqslant 0$ at any time t is represented by a spectrum which is a function of both direction and frequency, namely $F_x(\alpha,\beta,\omega)$, as

$$E_x(x,y,z,t) = \frac{1}{2\pi} \int_{-\infty}^{\infty} \int_{-\infty}^{\infty} \int_{-\infty}^{\infty} F_x(\alpha,\beta,\omega)$$

$$\exp\{j\,[\omega t - k(\alpha x + \beta y + \gamma z)]\}\, d\alpha d\beta d\omega \ , \qquad (3.73)$$

where

$$F_x(\alpha,\beta,\omega) = \frac{1}{\lambda^2} \int_{-\infty}^{\infty} \int_{-\infty}^{\infty} \int_{-\infty}^{\infty} E_{ax}(x,y,t)$$

$$\exp\{-j[\omega t - k(\alpha x + \beta y)]\}\, dx\, dy\, dt \ . \qquad (3.74)$$

3.5 COMPLETE REPRESENTATION OF THREE-DIMENSIONAL RADIATING FIELDS

3.5.1 Fields in a half space

Gathering together the results from earlier sections, starting with equation (3.16), if the tangential electric field over the aperture $(x - y)$ plane is

$$\mathbf{E}_a(x,y) = \mathbf{u}_x\, E_{ax}(x,y) + \mathbf{u}_y\, E_{ay}(x,y) \qquad (3.75)$$

then the field radiating into the half-space $z \geqslant 0$ is completely and uniquely represented by the two angular plane-wave spectra $F_x(\alpha,\beta)$ and $F_y(\alpha,\beta)$. They are, respectively,

$$F_x(\alpha,\beta) = \frac{1}{\lambda^2} \int_{-\infty}^{\infty} \int_{-\infty}^{\infty} E_{ax}(x,y) \exp\{jk\,(\alpha x + \beta y)\}\, dx\, dy \qquad (3.76)$$

and

$$F_y(\alpha,\beta) = \frac{1}{\lambda^2} \int_{-\infty}^{\infty} \int_{-\infty}^{\infty} E_{ay}(x,y) \exp\{jk\,(\alpha x + \beta y)\}\, dx\, dy \qquad (3.77)$$

The complete fields in $z \geqslant 0$ are then, from equations (3.11), (3.12) and (3.14), (3.15), in Cartesian components

$$\mathbf{E}(x,y,z) = \int_{-\infty}^{\infty} \int_{-\infty}^{\infty} [(\mathbf{u}_x - \mathbf{u}_z \frac{\alpha}{\gamma})\, F_x(\alpha,\beta) + (\mathbf{u}_y - \mathbf{u}_z \frac{\beta}{\gamma})\, F_y(\alpha,\beta)]$$

$$\exp(-jk\mathbf{u}.\mathbf{r}.)\, d\alpha d\beta \qquad (3.78)$$

where $\mathbf{u} = \mathbf{u}_x\alpha + \mathbf{u}_y\beta + \mathbf{u}_z\gamma$ and $\mathbf{r} = \mathbf{u}_x x + \mathbf{u}_y y + \mathbf{u}_z z$, and

$$\mathbf{H}(x,y,z) = \frac{1}{Z}\,\mathbf{u} \times \mathbf{E}(x,y,z) \tag{3.79}$$

These are exact, everywhere in $z \geqslant 0$.

Assuming the aperture field of equation (3.75) is of finite lateral extent, the radiated far-field ($kr \to \infty$) is asymptotically, in Cartesian-component form,

$$\mathbf{E}(x,y,z) \simeq j2\pi \frac{\exp(-jkr)}{kr} \left[(\mathbf{u}_x\,\gamma - \mathbf{u}_z\,\alpha)\,F_x(\alpha,\beta) + (\mathbf{u}_y\,\gamma - \mathbf{u}_z\,\beta)\,F_y(\alpha,\beta) \right] \tag{3.80}$$

where $r^2 = x^2 + y^2 + z^2$ and

$$\mathbf{H}(x,y,z) \simeq \frac{1}{Z}\,\mathbf{u} \times \mathbf{E}\,(x,y,z)\ . \tag{3.81}$$

Alternatively, in spherical-polar form, the far field is

$$\mathbf{E}(r,\theta,\phi) \simeq j2\pi \frac{\exp(-jkr)}{kr} \left[(\mathbf{u}_\theta\,\cos\phi - \mathbf{u}_\phi\,\cos\theta\,\sin\phi)\,F_x(\alpha,\beta) + (\mathbf{u}_\theta\,\sin\phi + \mathbf{u}_\phi\,\cos\theta\,\cos\phi)\,F_y(\alpha,\beta) \right] \tag{3.82}$$

in which $\alpha = \sin\theta\,\cos\phi$ and $\beta = \sin\theta\,\sin\phi$, and

$$\mathbf{H}(r,\theta,\phi) \simeq \frac{1}{Z}\,\mathbf{u}_r \times \mathbf{E}(r,\theta,\phi) \tag{3.83}$$

Note that the Rayleigh criterion (section 5.3.1) defines the approximate extent of the far-field region as

$$r \geqslant \frac{2D^2}{\lambda} \tag{3.84}$$

where D is the largest lateral dimension of the radiating aperture, which is that part of the aperture plane over which the tangential field is non-zero.

Exercise

Expand the vector products in equations (3.79), (3.81) and (3.83) to show that the magnetic fields in detail are

$$\mathbf{H}(x,y,z) = \frac{1}{Z}\int_{-\infty}^{\infty}\int_{-\infty}^{\infty} [(-\mathbf{u}_x \frac{\alpha\beta}{\gamma} + \mathbf{u}_y \frac{1-\beta^2}{\gamma} - \mathbf{u}_z\,\beta)\,F_x(\alpha,\beta)$$

$$+ (-\mathbf{u}_x \frac{1-\alpha^2}{\gamma} + \mathbf{u}_y \frac{\alpha\beta}{\gamma} + \mathbf{u}_z\,\alpha)\,F_y(\alpha,\beta)]$$

$$\exp(-jk\mathbf{u}\,.\,\mathbf{r})\,d\alpha d\beta \qquad (3.85)$$

everywhere in $z \geqslant 0$, and

$$\mathbf{H}(x,y,z) \simeq j\,\frac{2\pi}{Z}\,\frac{\exp(-jkr)}{kr}\left\{(-\mathbf{u}_x\,\alpha\beta + \mathbf{u}_y\,(1-\beta^2) - \mathbf{u}_z\,\beta\gamma]\,F_x(\alpha,\beta)\right.$$

$$\left. + [-\mathbf{u}_x\,(1-\alpha^2) + \mathbf{u}_y\,\alpha\beta + \mathbf{u}_z\,\alpha\gamma]\,F_y(\alpha,\beta)\right\} \qquad (3.86)$$

or

$$\mathbf{H}(r,\theta,\phi) \simeq j\,\frac{2\pi}{Z}\,\frac{\exp(-jkr)}{kr}\,[(\mathbf{u}_\theta\cos\theta\sin\phi + \mathbf{u}_\phi\cos\phi)\,F_x(\alpha,\beta)$$

$$+ (-\mathbf{u}_\theta\cos\theta\cos\phi + \mathbf{u}_\phi\sin\phi)\,F_y(\alpha,\beta)] \qquad (3.87)$$

in the far field as $kr \to \infty$.

3.5.2 Huygens' Principle

Collapsing the aperture down to an element $dxdy$ at the origin allows us to study the fields radiated by a Huygens element. This provides a tool for the investigation of some radiation problems, in particular those where the field is known over a non-planar surface. To this end we will eventually remove the Huygens element from the origin of the Cartesian coordinate system and give it an arbitrary orientation.

But first, suppose that the tangential electric field, at the element $dxdy$ at the origin, is

$$\mathbf{E}_a = \mathbf{u}_x\,E_{ax} + \mathbf{u}_y\,E_{ay} \qquad (3.88)$$

where the component fields E_{ax} and E_{ay} are of course constant. Since the extent of the aperture is necessarily much smaller than the wavelength, the angular spectra (equations (3.76) and (3.77)) are

$$F_x\,(\alpha,\beta) = \frac{E_{ax}}{\lambda^2}\;\mathrm{d}x\mathrm{d}y \tag{3.89}$$

and

$$F_y\,(\alpha,\beta) = \frac{E_{ay}}{\lambda^2}\;\mathrm{d}x\mathrm{d}y \tag{3.90}$$

which are independent of direction.

According to the Rayleigh far-field criterion (equation (3.84)) the far-field begins almost immediately. Hence the field due to the radiating element everywhere in the half space $z \geqslant 0$, except in the immediate vicinity of the element, is (from equations (3.80) and 3.81)), in Cartesian components,

$$\mathrm{d}\mathbf{E}(x,y,z) = j\,\frac{2\pi}{\lambda^2}\,\frac{\exp\,(-jkr)}{kr}\,[(\mathbf{u}_x\,\gamma - \mathbf{u}_z\,\alpha)\,E_{ax} + (\mathbf{u}_y\,\gamma - \mathbf{u}_z\,\beta)\,E_{ay}]\;\mathrm{d}x'\mathrm{d}y' \tag{3.91}$$

and

$$\mathrm{d}\mathbf{H}(x,y,z) = \frac{1}{Z}\;\mathbf{u}\times\mathrm{d}\mathbf{E}\,(x,y,z)\quad . \tag{3.92}$$

Primed coordinates will be used here to distinguish the source point from the observation point. In spherical-polar form the radiated field, except in the immediate vicinity of the source element, is (from equations (3.82) and (3.83))

$$\mathrm{d}\mathbf{E}(r,\theta,\phi) = j\,\frac{2\pi}{\lambda^2}\,\frac{\exp\,(-jkr)}{kr}\,[(\mathbf{u}_\theta\,\cos\phi - \mathbf{u}_\phi\,\cos\theta\,\sin\phi)\,E_{ax} + (\mathbf{u}_\theta\,\sin\phi + \mathbf{u}_\phi\,\cos\theta\,\cos\phi)\,E_{ay}]\;\mathrm{d}x'\mathrm{d}y' \tag{3.93}$$

and

$$\mathrm{d}\mathbf{H}(r,\theta,\phi) = \frac{1}{Z}\;\mathbf{u}_r\times\mathrm{d}\mathbf{E}\,(r,\theta,\phi)\quad . \tag{3.94}$$

In its Cartesian-component form (equation (3.91)) the vector part of the expression is

$$(\mathbf{u}_x \gamma - \mathbf{u}_z \alpha) E_{ax} + (\mathbf{u}_y \gamma - \mathbf{u}_z \beta) E_{ay} = \gamma \mathbf{E}_a - \mathbf{u}_z (\mathbf{u} \cdot \mathbf{E}_a) \tag{3.95}$$

where $\mathbf{E}_a$ is the tangential electric field at the element, as in equation (3.88). But equation (3.95) is also true if $\mathbf{E}_a$ is the total electric field at the element.

Exercise
Show that equation (3.95) is satisfied for

$$\mathbf{E}_a = \mathbf{u}_x E_{ax} + \mathbf{u}_y E_{ay} + \mathbf{u}_z E_{az} \quad .$$

Then the elemental radiated field, with the source element at the origin, becomes

$$d\mathbf{E}(x,y,z) = j \frac{2\pi}{\lambda^2} \frac{\exp(-jkr)}{kr} [\gamma \mathbf{E}_a - \mathbf{u}_z (\mathbf{u} \cdot \mathbf{E}_a)] \, dx'dy' \quad . \tag{3.96}$$

This can now be put in coordinate-free form, using the vector geometry of Fig. 3.11. Referred to the point O, the Huygens element is at position $\mathbf{r}'$ and the

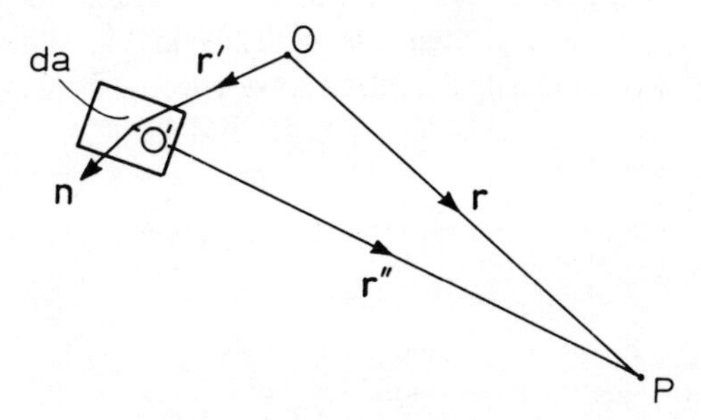

Fig. 3.11 – A Huygens element **n** *da* and a field point P.

field point is at position **r**. The element has area *da* and the direction of its normal is given by the unit vector **n**. The position of the field point P relative to the element at O′ is

$$\mathbf{r}'' = \mathbf{r} - \mathbf{r}' \quad . \tag{3.97}$$

Then, provided $\mathbf{n} \cdot \mathbf{u}_{r''}$ is positive, the field at P due to the element at O′ is

$$d\mathbf{E}(P) = j \frac{2\pi}{\lambda^2} \frac{\exp(-jkr'')}{kr''} [(\mathbf{u}_{r''} \cdot \mathbf{n}) \mathbf{E}_a - \mathbf{n} (\mathbf{u}_{r''} \cdot \mathbf{E}_a)] \, da \tag{3.98}$$

or, using equation (2.13),

$$\mathrm{d}\mathbf{E}(P) = j\,\frac{2\pi}{\lambda^2}\,\frac{\exp(-jkr'')}{kr''}\,\mathbf{u}_{r''} \times (\mathbf{E}_a \times \mathbf{n})\,\mathrm{d}a \tag{3.99}$$

in which $\mathbf{E}_a$ is either the total or the tangential field at the element. Note that this form of Huygens' Principle is exact, provided only that the observation point is more than a few wavelengths distant from the radiating element, and that $\mathbf{n}$ and $\mathbf{u}_{r''}$ have a positive scalar product.

In the limiting case, when the observation point P is infinitely far away, $\mathbf{r}''$ becomes parallel to $\mathbf{r}$, and the elemental field is

$$\underset{P\to\infty}{\mathrm{d}\mathbf{E}(P)} = j\,\frac{2\pi}{\lambda^2}\,\frac{\exp(-jkr)}{kr}\exp(+jk\,\mathbf{u}_r.\mathbf{r}')\,\mathbf{u}_r \times (\mathbf{E}_a \times \mathbf{n})\,\mathrm{d}a \tag{3.100}$$

in which

$$\mathbf{u}_r \times (\mathbf{E}_a \times \mathbf{n}) = (\mathbf{u}_r\,.\,\mathbf{n})\,\mathbf{E}_a - \mathbf{n}\,(\mathbf{u}_r\,.\,\mathbf{E}_a) \quad .$$

3.5.3 Fields in the whole space

The representation of fields radiated into the half-space $z \geqslant 0$ can be extended to the whole space, as indicated schematically in Fig. 3.12. We can envisage simply adding two more angular plane-wave spectra $F_x^-(\alpha,\beta)$ and $F_y^-(\alpha,\beta)$, for

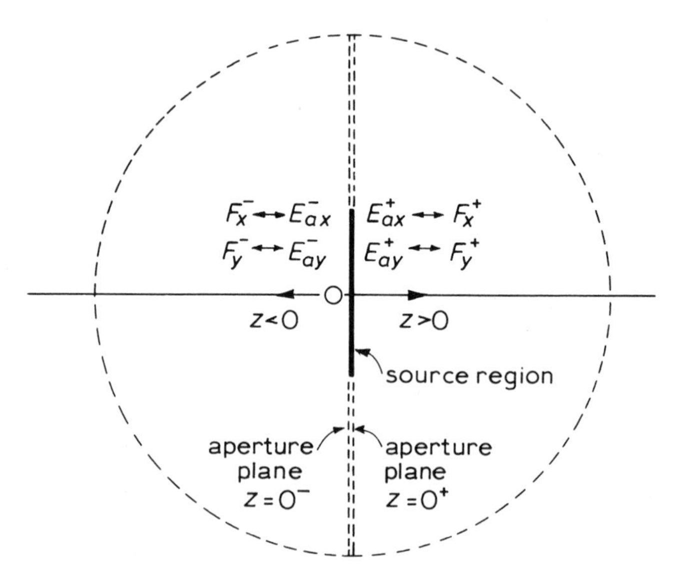

Fig. 3.12 – Scheme for representing fields radiated into the whole space.

the adjoining half space $z < 0$. These spectra are Fourier transforms of the tangential electric-field components $E_{ax}^-(x,y)$ and $E_{ay}^-(x,y)$ over the plane $z = 0^-$. The spectra for the original half space $z > 0$ should now be written, for consistency, as $F_x^+(\alpha,\beta)$ and $F_y^+(\alpha,\beta)$ being the Fourier transforms of the tangential electric-field components $E_{ax}^+(x,y)$ and $E_{ay}^+(x,y)$ over the plane $z = 0^+$. Then the far field is given in any direction $\mathbf{u}_r$, asymptotically as $kr \to \infty$, by

$$\mathbf{E}(r,\theta,\phi) \simeq \frac{\exp(-jkr)}{kr}\,\mathbf{e}(\theta,\phi) \tag{3.101}$$

and

$$\mathbf{H}(r,\theta,\phi) \simeq \frac{1}{Z}\,\mathbf{u}_r \times \mathbf{E}(r,\theta,\phi)\ ;\ \mathbf{u}_r \,.\, \mathbf{E}(r,\theta,\phi) = 0 \tag{3.102}$$

with

$$\begin{aligned}\mathbf{e}(\theta,\phi) = j\,2\pi\,\{&\mathbf{u}_\theta\,[F_x\,(\alpha,\beta)\cos\phi + F_y\,(\alpha,\beta)\sin\phi] \\ &+ \mathbf{u}_\phi\,[-F_x\,(\alpha,\beta)\sin\phi + F_y\,(\alpha,\beta)\cos\phi]\,\cos\theta\}\end{aligned} \tag{3.103}$$

in which

$$F_x\,(\alpha,\beta) = \begin{cases} F_x^+\,(\alpha,\beta) & \text{for } \mathbf{u}_r \,.\, \mathbf{u}_z > 0 \\ F_x^-\,(\alpha,\beta) & \text{for } \mathbf{u}_r \,.\, \mathbf{u}_z < 0 \end{cases} \tag{3.104}$$

and

$$F_y\,(\alpha,\beta) = \begin{cases} F_y^+\,(\alpha,\beta) & \text{for } \mathbf{u}_r \,.\, \mathbf{u}_z > 0 \\ F_y^-\,(\alpha,\beta) & \text{for } \mathbf{u}_r \,.\, \mathbf{u}_z < 0 \end{cases} \tag{3.105}$$

where

$$F_x^\pm\,(\alpha,\beta) \longleftrightarrow E_{ax}^\pm\,(x,y) \tag{3.106}$$

and

$$F_y^\pm\,(\alpha,\beta) \longleftrightarrow E_{ay}^\pm\,(x,y)\ .$$

For convenience we have supposed that the spherical-polar coordinates, aperture-plane coordinates, and direction cosines for the left half space $z < 0$ are designed to look the same when viewed through the aperture from the right as those for the right half plane when they are viewed through the aperture from the opposite side.

Example of radiation from an elementary (Hertzian) dipole.
This is one of the few antennas that precisely fits the scheme of Fig. 3.12. It consists (see Fig. 3.13) of an infinitely thin, perfectly conducting strip carrying a current I and whose dimensions a and b are both very small in comparison

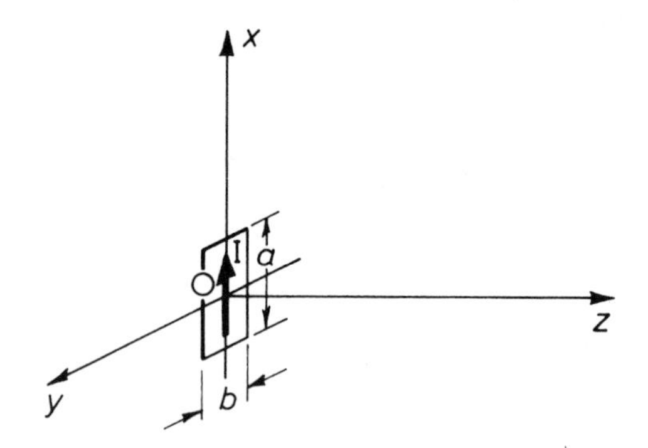

Fig. 3.13 – A radiating elementary dipole strip.

with the wavelength. Then the aperture field is best prescribed in this case in terms of tangential magnetic field. Except on the strip the aperture field is zero since, by symmetry, $\mathbf{H}_{tan} \equiv 0$. Ampere's law (see Appendix B 1.4) gives the tangential magnetic field on the surface of the strip. Hence the aperture field over the $z = 0^+$ plane is

$$H_{ay}^{+}(x,y) = -\frac{I}{2b}\,\text{rect}_a(x)\,\text{rect}_b(y) \quad , \tag{3.107}$$

where $\text{rect}_a(x)$ is the **rectangle function**, which is unity for $-a/2 < x < a/2$ and is zero elsewhere. A similar expression give the aperture field $H_{ay}^{-}(x,y)$ over the $z = 0^-$ plane.

In order to proceed we need the dual of the representation in section 3.5.1, which is for tangential-electric aperture fields. Using the Principle of Duality (Appendix B 2.7) we have, for a y-polarized magnetic aperture field $H_{ay}(x,y)$ (that is, $H_{ax} \equiv 0$),

$$\mathbf{H}(x,y,z) = \int_{-\infty}^{\infty}\int_{-\infty}^{\infty} (\mathbf{u}_y - \mathbf{u}_z \frac{\beta}{\gamma})\, G_y(\alpha,\beta)$$

$$\exp\{-jk\,(\alpha x + \beta y + \gamma z)\}\, d\alpha d\beta \tag{3.108}$$

and

$$\mathbf{E}(x,y,z) = -Z\,\mathbf{u} \times \mathbf{H}\,(x,y,z)$$

$$= Z \int_{-\infty}^{\infty} \int_{-\infty}^{\infty} (\mathbf{u}_x \frac{1-\alpha^2}{\gamma} - \mathbf{u}_y \frac{\alpha\beta}{\gamma} - \mathbf{u}_z\,\alpha)\, G_y(\alpha,\beta)$$

$$\exp\,\{-jk\,(\alpha x + \beta y + \gamma z)\}\; \mathrm{d}\alpha \mathrm{d}\beta\;. \qquad (3.109)$$

The y-polarized magnetic angular spectrum is therefore

$$G_y(\alpha,\beta) = \frac{1}{\lambda^2} \int_{-\infty}^{\infty} \int_{-\infty}^{\infty} H_{ay}(x,y) \exp\,\{jk\,(\alpha x + \beta y)\}\; \mathrm{d}x\, \mathrm{d}y \qquad (3.110)$$

and the far field is (for $kr \to \infty$)

$$\mathbf{E}(r,\theta,\phi) \simeq j\,2\pi Z \frac{\exp\,(-jkr)}{kr} G_y(\alpha,\beta)\; [\mathbf{u}_x\,(1-\alpha^2) - \mathbf{u}_y\,\alpha\beta - \mathbf{u}_z\,\alpha\gamma] \qquad (3.111)$$

with

$$\mathbf{H}(r,\theta,\phi) \simeq \frac{1}{Z}\,\mathbf{u}_r \times \mathbf{E}(r,\theta,\phi)\;\;. \qquad (3.112)$$

Substituting equation (3.107) into (3.110), the magnetic angular spectrum in the positive half space is

$$G_y^+\,(\alpha,\beta) = -\frac{Ia}{2\lambda^2}\;\;, \qquad (3.113)$$

for vanishingly small a and b. The electric far-field, which begins at a few wavelengths distance from the dipole, is

$$\mathbf{E}(r,\theta,\phi) = j\,2\pi\, \frac{\exp\,(-jkr)}{kr} \frac{ZIa}{2\lambda^2}\; [-\mathbf{u}_x\,(1-\alpha^2) + \mathbf{u}_y\,\alpha\beta + \mathbf{u}_z\,\alpha\gamma] \qquad (3.114)$$

with the magnetic far field given by equation (3.112).

The conventional representation of the radiation from an elementary dipole is with the current element Ia directed along the z-axis, instead of the x-axis as in Fig. 3.13. Making the appropriate transformation (which is done by recognising that the direction cosines remain unchanged but are simply assigned with respect to different axes) the far field becomes, in the conventional dipole coordinate system,

$$\mathbf{E}(r,\theta,\phi) = \mathbf{u}_\theta \, j \, 2\pi \frac{\exp(-jkr)}{kr} \frac{ZIa}{2\lambda^2} \sin\theta \tag{3.115}$$

and

$$H_\phi = \frac{E_\theta}{Z} \; . \tag{3.116}$$

The above analysis can be repeated for the half space $z < 0$, giving precisely the same fields.

The representation of fields in the whole space by means of two pairs of angular spectra, one pair for each half space, is fortunately not confined to infinitely thin antennas. As will be shown in the first few pages of Chapter 5, the choice of the position of the aperture plane is arbitrary. So, provided they are clear of the structure of the antenna, the two aperture planes can be positioned immediately to the right and left of the antenna. The fields are thereby specified in almost the whole space. Indeed the angular spectra based on a knowledge of the tangential fields over these actual aperture planes can be transformed back on to the plane $z = 0$ to give 'equivalent aperture fields' from which the fields in the whole space, except in the interior and interstices of the antenna itself, can be derived.

CHAPTER 4

Transmitting and receiving antennas

We have seen in previous chapters how the radiated field can be described in terms of an angular spectrum of plane waves, and how the spectrum function is directly related to the far-field pattern of the radiation. This notion will be applied, in general terms in this chapter, to the field radiated by a transmitting antenna. It will then be shown, using the reciprocity theorem, how the response of an antenna, when used as a receiver and illuminated by a plane wave, is related to the angular spectrum when the antenna is used as a transmitter. Finally, since the field radiated by a transmitting antenna is a spectrum of plane waves, it simply remains to integrate the receiving-antenna response over all directions in order to obtain a formula for the field coupled from one antenna into another. This antenna coupling formula applies for arbitrary separations between transmitter and receiver. It can therefore be used as a basis for the (radiating) near-field antenna measurement technique, as well as yielding the Friis far-field coupling formula for infinitely large separations.

4.1 THE ANTENNA AS TRANSMITTER

At the end of the last chapter it was established that the electromagnetic field at a large distance r from a radiating aperture of finite size, such as an antenna, has the approximate form

$$\mathbf{E}(r,\theta,\phi) = \frac{\exp(-jkr)}{kr}\,\mathbf{e}(\theta,\phi) \tag{4.1}$$

and

$$\mathbf{H}(r,\theta,\phi) = \frac{1}{Z}\,\mathbf{u}_r \times \mathbf{E}(r,\theta,\phi) \tag{4.2}$$

with

$$\mathbf{u}_r \,.\, \mathbf{e}(\theta,\phi) = 0 \tag{4.3}$$

the approximation tending to an equality as $kr \to \infty$. It is known as the far field. The geometry is that shown in Fig. 4.1. The field point P has spherical-polar coordinates (r,θ,ϕ), and $\mathbf{u}_r$ is a unit vector in the direction OP. All directions are included, and the polar angle θ and azimuth angle ϕ have the ranges $-\pi \leqslant \theta < \pi$ and $-\pi/2 \leqslant \phi < \pi/2$. The far field described by equations (4.1), (4.2) and (4.3) is a spherical wave whose electric field is given in a particular direction by the vector pattern function $\mathbf{e}(\theta,\phi)$. In the locality of a point such as P the fields have the approximate form of a plane wave travelling in the direction $\mathbf{u}_r$, with $\mathbf{E}$ and $\mathbf{H}$ mutually at right angles and tangential to the sphere of radius r. There is no radial field component. The constants k and Z are the phase constant and plane-wave impedance of the medium. The time dependence $\exp(j\omega t)$ is suppressed.

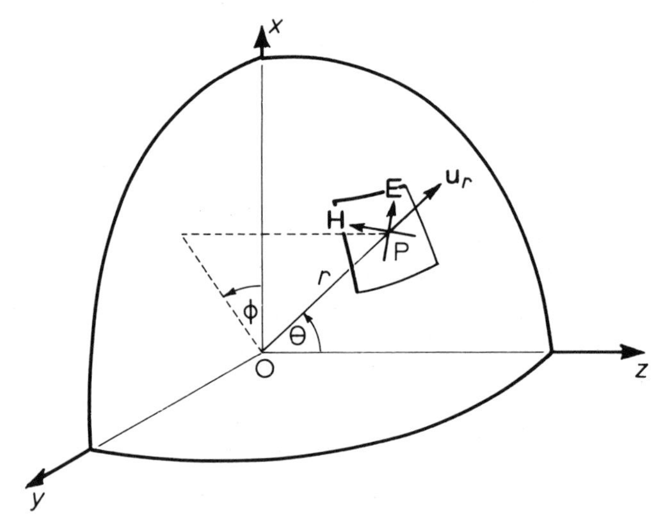

Fig. 4.1 – Geometry for describing the transmitted far field.

Poynting's theorem (equation (2.10)) gives the power flow at P as being purely radial, and having the value

$$S_r(r,\theta,\phi) = \frac{|e(\theta,\phi)|^2}{2Z(kr)^2} \tag{4.4}$$

which is obtained by substituting equations (4.1) and (4.2) into equation (2.10). If we assume that the radiated field of equations (4.1) to (4.3) arises when power P_o is delivered to the antenna, the definition of section 1.4 gives the gain function of an antenna as

$$G(\theta,\phi) = \frac{4\pi r^2}{P_o} S_r(r,\theta,\phi) \tag{4.5}$$

or

$$G(\theta,\phi) = \frac{\lambda^2 |e(\theta,\phi)|^2}{2\pi Z P_o} \tag{4.6}$$

If we make the further assumption that the antenna is 100% efficient, which means that all the power P_o delivered to the antenna is radiated, the gain function $G(\theta,\phi)$ and the directivity function $D(\theta,\phi)$ are identical. The gain of an antenna is usually taken to be the maximum value of the gain function, namely,

$$G_o = [G(\theta,\phi)]_{max} \tag{4.7}$$

when this maximum can be identified unambiguously.

4.1.1 Gain function and the angular spectrum

The general formula for the vector pattern function $\mathbf{e}(\theta,\phi)$ in the angular plane-wave spectrum representation is given by equation (3.103). Four angular spectra are required: two for each polarization of the aperture field for each half-space. In order to simplify the discussion, both now and later, we will confine our attention to the half space $z \geqslant 0$ and to the single polarization of an x-directed aperture-field distribution $E_{ax}(x,y)$. Then the only angular spectrum function involved is $F_x(\alpha,\beta)$ of equation (3.76), which is

$$F_x(\alpha,\beta) = \frac{1}{\lambda^2} \int_{-\infty}^{\infty} \int_{-\infty}^{\infty} E_{ax}(x,y) \exp \{jk(\alpha x + \beta y)\}\, dx dy \tag{4.8}$$

in terms of which the far-field vector pattern function is

$$\mathbf{e}(\theta,\phi) = j2\pi F_x(\sin\theta \cos\phi, \sin\theta \sin\phi)$$

$$[\mathbf{u}_\theta \cos\phi - \mathbf{u}_\phi \cos\theta \sin\phi] \tag{4.9}$$

Hence the gain function for an x-polarized aperture field in a direction (θ,ϕ) pointing into the half space $z \geqslant 0$ is

$$G(\theta,\phi) = \frac{2\pi\lambda^2}{ZP_o} |F_x(\sin\theta\cos\phi, \sin\theta\sin\phi)|^2 (1 - \sin^2\theta \sin^2\phi) \tag{4.10}$$

where P_o is the power delivered to the antenna.

For a narrow-beam pattern whose maximum is in the direction $\theta = 0$,

$$G(\theta,\phi) \approx \frac{2\pi\lambda^2}{ZP_o} |F_x|^2 \tag{4.11}$$

in which the functional dependence of F_x on θ and ϕ has been suppressed for brevity.

These results can easily be extended to the general case of a pattern function described by equation (3.103). In this case

$$\mathbf{e}(\theta,\phi) = j2\pi\ [\mathbf{u}_\theta\ (F_x \cos\phi + F_y \sin\phi) + \mathbf{u}_\phi\ (-F_x \cos\theta \sin\phi + F_y \cos\theta \cos\phi)] \tag{4.12}$$

where F_x and F_y are the angular spectra $F_x^{\pm}(\alpha,\beta)$, the + sign referring to directions (θ,ϕ) pointing into the half space $z \geqslant 0$, the negative sign referring to the half space $z \leqslant 0$. The gain function is then given by

$$G(\theta,\phi) = \frac{2\pi\lambda^2}{ZP_o} \left\{ |F_x|^2(1 - \sin^2\theta \sin^2\phi) + 2\ \mathrm{Re}\ [F_x F_y^*]\ \sin^2\theta \sin\phi \cos\phi + |F_y|^2\ (1 - \sin^2\theta \cos^2\phi) \right\} \tag{4.13}$$

for total power P_o delivered to the antenna. For patterns narrowly centred on the direction $\theta = 0$,

$$G(\theta,\phi) \approx \frac{2\pi\lambda^2}{ZP_o} \left\{ |F_x|^2 + |F_y|^2 \right\} \quad . \tag{4.14}$$

Exercise

Derive equation (4.13).

Exercise

Find the gain function for an antenna whose aperture-field distribution is that of equation (3.26), which is x-polarized, uniform in amplitude in one direction and half-cosine in the other, and entirely equiphase over the rectangular aperture $a \times b$. Show that the gain is

$$G_o = \frac{32ab}{\pi\lambda^2} \quad . \tag{4.15}$$

Assume that all the power delivered to the antenna is radiated.

4.1.2 Gain of a uniformly illuminated aperture

A useful standard of comparison in antenna design is the radiation from a uniformly illuminated aperture. It transpires that the gain of such an aperture field distribution is easily calculated and is also the maximum obtainable from an aperture of that area.

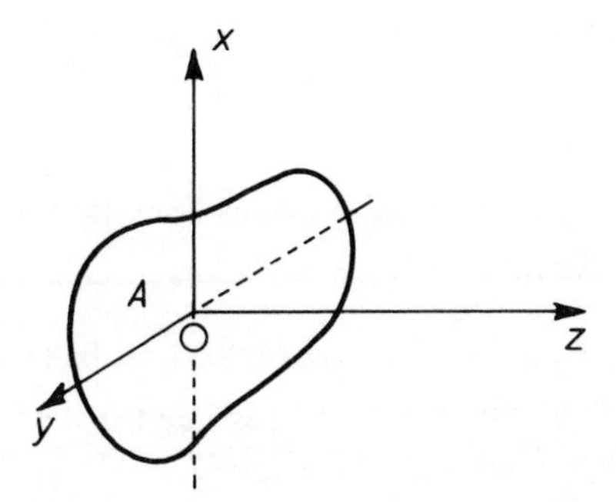

Fig. 4.2 – Aperture of arbitrary shape and area A.

Suppose that the aperture is that shown in Fig. 4.2, of arbitrary shape and area A, lying in the x-y plane. Let the aperture field be the constant E_o over A and zero outside, that is,

$$\begin{aligned} E_{ax}(x,y) &= E_o \text{ over } A \\ &= 0 \quad \text{outside } A \ . \end{aligned} \tag{4.16}$$

The angular spectrum for this aperture field is

$$F_x(\alpha,\beta) = \frac{E_o}{\lambda^2} \iint_A \exp\{jk(\alpha x + \beta y)\, \mathrm{d}x\mathrm{d}y \tag{4.17}$$

which is clearly a maximum in the direction of the z-axis ($\theta = 0$ and $\alpha = \beta = 0$), since it is the only direction in which all the contributions from the integrand are in phase. Then

$$F_x(0,0) = \frac{E_o}{\lambda^2} \iint_A \mathrm{d}x\mathrm{d}y = \frac{E_o A}{\lambda^2} \ . \tag{4.18}$$

The total power passing through the aperture is

$$P_o = \frac{|E_o|^2 A}{2Z} \tag{4.19}$$

and will be assumed to be equal to that delivered to the antenna. Then equations (4.7), (4.10), (4.18) and (4.19) give the gain G_o of a uniformly illuminated aperture of area A as

$$G_o = \frac{4\pi A}{\lambda^2} \quad . \tag{4.20}$$

Exercise

Check that the gain obtained in the previous Exercise for a non-uniform aperture field is less than if it had been uniform.

To show that the gain of equation (4.20) is in fact the maximum obtainable (except in abnormal circumstances of 'super gain') from an aperture of area A, let the aperture field $E_{ax}(x,y)$ be arbitrary over A, and zero outside, with maximum modulus

$$\max |E_{ax}(x,y)| = |E_o| \quad . \tag{4.21}$$

The gain for this arbitrary aperture field is

$$G_o = \frac{4\pi}{\lambda^2} \frac{\left|\iint_A E_{ax}(x,y)\,\mathrm{d}x\mathrm{d}y\right|^2}{\iint_A |E_{ax}(x,y)|^2\,\mathrm{d}x\mathrm{d}y} \quad . \tag{4.22}$$

Now it is clear physically, and can be proved using Schwarz's inequality, that

$$\begin{aligned} \left|\iint_A E_{ax}(x,y)\mathrm{d}x\,\mathrm{d}y\right|^2 &\leqslant \iint_A |E_{ax}(x,y)|^2\,\mathrm{d}x\,\mathrm{d}y \iint_A \mathrm{d}x\,\mathrm{d}y \\ &\leqslant \iint_A |E_o|^2\,\mathrm{d}x\,\mathrm{d}y \iint_A \mathrm{d}x\mathrm{d}y \end{aligned} \tag{4.23}$$

and hence, in general for an aperture of area A, that

$$G_o \leqslant \frac{4\pi A}{\lambda^2} \tag{4.24}$$

the equality holding if, and only if, $E_{ax}(x,y)$ is constant in both amplitude and phase, as for equation (4.20).

Exercise
Find the gain functions for uniformly illuminated rectangular and circular apertures, and hence show that equation (4.20) is satisfied in those two cases.

4.1.3 Polarization and gain function

The polarization of the far field transmitted by an antenna is the direction of the pattern function vector $\mathbf{e}(\theta,\phi)$, which changes as the direction of observation (θ,ϕ) changes, but is always tangential to the sphere of observation of radius r. In the case of an x-polarized aperture field the vector pattern function in the half space $z \geqslant 0$ is

$$\mathbf{e}(\theta,\phi) = j2\pi F_x (\sin\theta\cos\phi, \sin\theta\sin\phi)$$

$$[\mathbf{u}_\theta \cos\phi - \mathbf{u}_\phi \cos\theta\sin\phi] \quad . \qquad (4.9)$$

This clearly consists of the sum of two orthogonal, linearly polarized waves in the perpendicular directions $\mathbf{u}_\theta$ and $\mathbf{u}_\phi$ (see Fig. 4.3), their complex amplitudes being in the ratio $-\cos\theta\tan\phi$. The polarization in this case is therefore always linear, but directed differently in different directions. If a receiving antenna

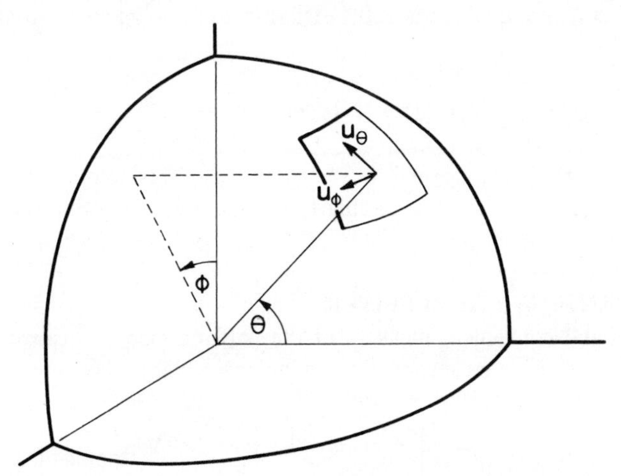

Fig. 4.3 – Directions of the unit vectors $\mathbf{u}_\theta$ and $\mathbf{u}_\phi$ on the sphere of observation.

which responds only to linear polarization were used to probe this field it would have to be reoriented to the vector direction of $\mathbf{e}(\theta,\phi)$ at each point (θ,ϕ) on the sphere of observation in order to register the gain function of equation (4.6). Looked at another way, since the field $\mathbf{e}(\theta,\phi)$ can be resolved into the sum of any two orthogonally polarized waves (see section 2.1 (f)) the gain function can likewise be resolved into the sum of two corresponding gain functions, as the following argument shows.

Write the vector pattern function $\mathbf{e}(\theta,\phi)$ for a particular direction as the weighted sum of two orthonormal waves $\mathbf{e}_1$ and $\mathbf{e}_2$, namely as

$$\mathbf{e} = a_1 \mathbf{e}_1 + a_2 \mathbf{e}_2 \tag{4.25}$$

where a_1 and a_2 are complex scalar amplitudes. For $\mathbf{e}_1$ and $\mathbf{e}_2$ to be **orthonormal** means that they must satisfy both the condition for being orthogonal, that is,

$$\mathbf{e}_1 \,.\, \mathbf{e}_2^* = 0 \tag{4.26}$$

and the condition that they are normalized, that is,

$$\mathbf{e}_1 \,.\, \mathbf{e}_1^* = \mathbf{e}_2 \,.\, \mathbf{e}_2^* = 1 \quad . \tag{4.27}$$

It follows from equations (4.25) – (4.27) that

$$|e|^2 = \mathbf{e} \,.\, \mathbf{e}^* = |a_1|^2 + |a_2|^2 \tag{4.28}$$

and hence that the gain function of equation (4.6), which is identical to directivity for an antenna which is 100% efficient, can also be resolved into two parts, namely

$$G(\theta,\phi) = G_1(\theta,\phi) + G_2(\theta,\phi) \tag{4.29}$$

in terms of the orthonormal waves $\mathbf{e}_1$ and $\mathbf{e}_2$. This resolution is particularly useful when $\mathbf{e}_1$ and $\mathbf{e}_2$ are two circularly polarized waves of opposite senses.

4.2 THE ANTENNA AS RECEIVER

A natural and convenient measure of the performance of an antenna when it is used as a receiver is its response to an incoming plane wave. Fig. 4.4 shows such

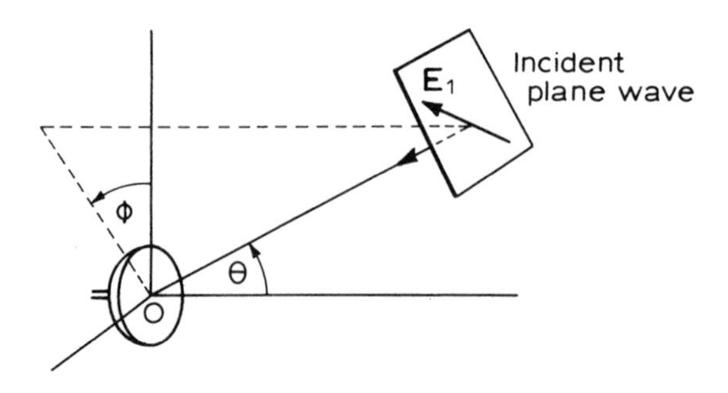

Fig. 4.4 – Plane wave incident on a receiving antenna

a plane wave $\mathbf{E}_1$ arriving from a direction (θ,ϕ) as defined in the antenna's coordinate system. The form this response takes (see section 1.4) is that of the ratio of the power delivered to the antenna divided by the power density of the incoming plane wave. It therefore has the dimensions of area and is a function of the direction of arrival. It is known as the effective receiving area $A(\theta,\phi)$ of the antenna. It is assumed that the polarization of the incident plane wave is adjusted in every direction (θ,ϕ) to deliver maximum power to the receiver; a condition known as **polarization match.**

Intuition suggests, and it will be proved in the following subsection, that the condition of polarization match is fulfilled when the polarization of the incident plane wave $\mathbf{E}_1$ is the same as that of the vector pattern function $\mathbf{e}(\theta,\phi)$ on transmission, and further that the antenna's effective receiving area $A(\theta,\phi)$ and its gain function $G(\theta,\phi)$ are proportional to one another. The precise relationship will shortly be shown to be

$$G(\theta,\phi) = \frac{4\pi}{\lambda^2} A(\theta,\phi) \quad . \tag{4.30}$$

This relationship is of the same form as equation (4.20), thereby presenting a coherent physical picture. Before going on to give a proof of equation (4.30) we will give an example of its application.

The Friis (Far-field) Transmission Formula

A receiving antenna at O′, Fig. 4.5, is in the far field of a transmitting antenna at O. If power P_T is fed to the transmitter, the power density at the receiving antenna, a distance r away in the direction (θ,ϕ) from the transmitter, will be $P_T G_T(\theta,\phi)/4\pi r^2$

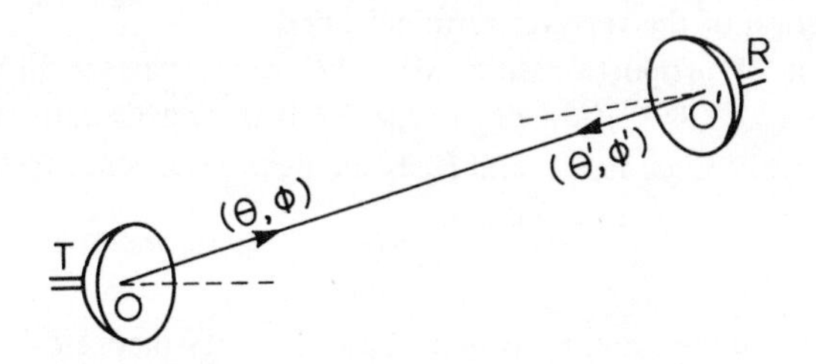

Fig. 4.5 – T transmitting to a receiver R.

in the approximate form of a plane wave. $G_T(\theta,\phi)$ is the gain function of the transmitter. If the effective receiving area of the receiving antenna is $A_R(\theta',\phi')$ in the direction (θ',ϕ') towards the transmitter, the power delivered to the receiver will be

$$P_\mathrm{R} = \frac{P_\mathrm{T}\, G_\mathrm{T}(\theta,\phi)\, A_\mathrm{R}(\theta',\phi')}{4\pi r^2} \quad . \tag{4.31}$$

Equation (4.30) may be used to replace the effective receiving area by the gain function $G_R(\theta',\phi')$ (on transmission) of the receiving antenna, so that

$$P_R = \frac{\lambda^2 P_T \, G_T(\theta,\phi) \, G_R(\theta',\phi')}{(4\pi r)^2} \qquad (4.32)$$

a formula that was originally developed by Friis (1937).

A general formula for the coupling between two antennas, not restricted to when the receiving antenna is in the far field of the transmitter, will be developed in section 4.3. Its development requires a precise statement of the relationship between the transmitting and receiving properties of an antenna, which is given in the following theorem.

4.2.1 Antenna reciprocity theorem

Theorem:

If a plane wave, whose electric field has the vector amplitude $\mathbf{e}_i$, is incident on an antenna from the direction $\mathbf{u}_i$ (as seen from the antenna) then the complex amplitude of the signal delivered by the antenna to the receiver is proportional to the scalar product $\mathbf{e}_i \, . \, \mathbf{e}(\mathbf{u}_i)$, where $\mathbf{e}(\mathbf{u}_i)$ is the vector pattern function of the field in the direction $\mathbf{u}_i$ when the antenna is used as a transmitter.

Note that:

(a) the antenna is assumed to be made from linear, reciprocal components,

(b) the signal is measured at some point in the antenna feeder and is expressed as a ratio of the signal at the same point on transmission,

(c) the feed is assumed to be matched, both in the direction of the antenna and in the direction of the receiver terminals, and

(d) the constant of proportionality, with the vector pattern function defined by equation (4.1), is $\lambda^2/(j4\pi \, ZP_o)$, where λ is the wavelength in the medium, Z its plane-wave impedance, and P_o is the power delivered to the antenna to give rise to the radiated field $\mathbf{e}(\mathbf{u}_i)$.

Proof:

The proof is based on the Lorentz form of the reciprocity theorem (see Appendix B section B.2.8). In a source-free region bounded by the closed surface A, if $(\mathbf{E}_1,\mathbf{H}_1)$ and $(\mathbf{E}_2,\mathbf{H}_2)$ are two independent sinusoidal field solutions of the same frequency, they are related to each other on the surface A by the closed-surface integral

$$\oint_A (\mathbf{E}_1 \times \mathbf{H}_2 - \mathbf{E}_2 \times \mathbf{H}_1) \, . \, \mathbf{n} \, \mathrm{d}A = 0 \quad , \qquad (4.33)$$

where the unit vector $\mathbf{n}$ is the inward normal to the surface A, and $\mathrm{d}A$ is an element of surface area.

The fields $(\mathbf{E}_1,\mathbf{H}_1)$ will be taken to be those associated with the antenna when it is transmitting, and the fields $(\mathbf{E}_2,\mathbf{H}_2)$ those associated with the antenna when it is receiving. The closed surface A is shown in Fig. 4.6 for a horn antenna connected by a metal-walled waveguide to a completely screened housing for the source/receiver. (It will become obvious that the proof is not restricted to horn antennas. Indeed the proof applies to any well screened antenna.) The surface A consists of a sphere of arbitrarily large radius r centred on the point O in the aperture of the antenna. The surface is indented at a convenient point to cover the whole surface of the source/receiver housing, the feeder waveguide and the antenna itself. The surface A is completed by spanning the cross-section of the feeder waveguide well away from the antenna structure.

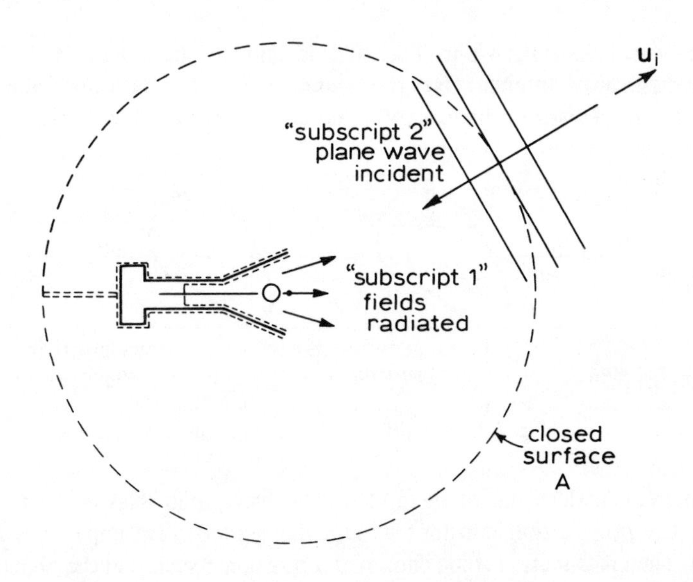

Fig. 4.6 – The surface A constructed to enclose a source-free volume in order to apply the Lorentz reciprocity theorem to fields radiated and received by an antenna.

There will be no contribution to the integral of equation (4.33) from the metal surface of the antenna, its feed and the source/receiver housing, since the tangential electric fields are zero there. (This will be so even if the metal is imperfectly conducting, provided that the thickness of the metal is significantly greater than the skin depth.) The parts of the closed surface A that remain, from which there will be a contribution to the integral of equation (4.33), are the cross-section A_g of the feeder waveguide and the entire surface A_r of the sphere of radius r. These two contributions will be considered separately.

Over the waveguide cross-section A_g
The transverse fields in the waveguide (see section A.10 of Appendix A) are, on transmission (subscript 1),

$$\mathbf{E}_1(x,y,z) = a\,\mathbf{e}_g(x,y)\exp(-j\beta_g z)$$

and

$$\mathbf{H}_1(x,y,z) = \frac{1}{Z_g}\,\mathbf{u}_z \times \mathbf{E}_1(x,y,z) \tag{4.34}$$

with

$$\mathbf{u}_z \,.\, \mathbf{E}_1(x,y,z) = 0 \quad .$$

The waveguide, shown in Fig. 4.7, is uniform in cross-section, lossless and matched to the antenna. Over the reference plane $z = 0$ the normalized transverse electric field distribution $\mathbf{e}_g\,(x,y)$ is assumed to be real, but is otherwise arbitrary,

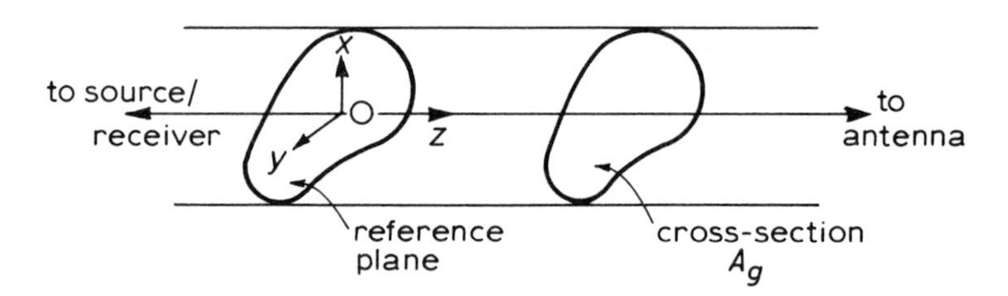

Fig. 4.7 – Coordinate system in the feeder waveguide.

consistent with being single-moded. The waveguide is assumed to be lossless, and the propagation constant β_g and the guided wave impedance Z_g are both real. The cross-section A_g is chosen at a position z such that the 'signal' amplitude a of the transmitted wave is real. Then, if the power delivered to the antenna is P_o, by Poynting's theorem

$$\tfrac{1}{2}\,\mathrm{Re}\iint_{A_g} \mathbf{E}_1(x,y,0) \times \mathbf{H}_1{}^*(x,y,0)\,\mathrm{d}x\mathrm{d}y$$

$$= \frac{a^2}{2Z_g}\iint_{A_g} e_g^2(x,y)\,\mathrm{d}x\mathrm{d}y = P_o \tag{4.35}$$

since both a and $\mathbf{e}_g\,(x,y)$ have been chosen to be real.

On reception (subscript 2), the transverse fields in the waveguide at the cross-section A_g are

$$\mathbf{E}_2(x,y,z) = b\,\mathbf{e}_g(x,y)\exp(+j\beta_g z)$$

and

$$\mathbf{H}_2(x,y,z) = -\frac{1}{Z_g}\mathbf{u}_z \times \mathbf{E}_2(x,y,z) \tag{4.36}$$

with

$$\mathbf{u}_z \,.\, \mathbf{E}_2(x,y,z) = 0$$

which is in the same mode as the waveguide field on transmission, is matched to the receiver, and the signal amplitude b must be taken in general to be complex.

The contribution to the Lorentz integral of equation (4.33) from the waveguide cross-section A_g is therefore

$$\iint_{A_g} (\mathbf{E}_1 \times \mathbf{H}_2 - \mathbf{E}_2 \times \mathbf{H}_1)\,.\,\mathbf{n}\,\mathrm{d}A$$

$$= \iint_{A_g} [\mathbf{E}_1(x,y,0) \times \mathbf{H}_2(x,y,0) - \mathbf{E}_2(x,y,0) \times \mathbf{H}_1(x,y,0)]\,.\,\mathbf{u}_z\,\mathrm{d}x\mathrm{d}y$$

$$= \frac{2ab}{Z_g}\iint_{A_g} e_g^2(x,y)\,\mathrm{d}x\mathrm{d}y = 4P_\mathrm{o}\frac{b}{a} \tag{4.37}$$

from equations (4.34) and (4.36) with equation (4.35) substituted. Note that it is in the feeder waveguide that the signal a on transmission and the signal b on reception become related by the Lorentz integral.

Over the sphere A_r

The field on transmission (subscript 1) is the spherical wave of equations (4.1) to (4.3) which are asymptotically as $kr \to \infty$

$$\mathbf{E}_1(r,\theta,\phi) = \frac{\exp(-jkr)}{kr}\,\mathbf{e}(\theta,\phi)$$

and

$$\mathbf{H}_1(r,\theta,\phi) = \frac{1}{Z}\,\mathbf{u}_r \times \mathbf{E}_1(r,\theta,\phi) \tag{4.38}$$

with

$$\mathbf{u}_r \,.\, \mathbf{E}_1(r,\theta,\phi) = 0 \quad .$$

The field on reception (subscript 2) at the point r (spherical coordinates r,θ,ϕ) due to the plane wave travelling in the direction $-\mathbf{u}_i$, whose vector electric field is $\mathbf{e}_i$, is

$$\mathbf{E}_2(r,\theta,\phi) = \mathbf{e}_i \exp(+jkr\,\mathbf{u}_i \,.\, \mathbf{u}_r)$$

and

$$\mathbf{H}_2(r,\theta,\phi) = -\frac{1}{Z}\,\mathbf{u}_i \times \mathbf{E}_2(r,\theta,\phi) \tag{4.39}$$

with

$$\mathbf{u}_i \,.\, \mathbf{E}_2(r,\theta,\phi) = 0 \quad .$$

Hence the contribution over the sphere of radius r to the Lorentz integral of equation (4.33) is

$$\iint_{A_r} (\mathbf{E}_1 \times \mathbf{H}_2 - \mathbf{E}_2 \times \mathbf{H}_1) \,.\, \mathbf{n}\, \mathrm{d}A$$

$$= \frac{\exp(-jkr)}{Zkr} \int_{-\pi/2}^{\pi/2} \int_{-\pi}^{\pi} \Big\{ \mathbf{e}(\theta,\phi) \times [(-\mathbf{u}_i) \times \mathbf{e}_i] - \mathbf{e}_i \times [\mathbf{u}_r \times \mathbf{e}(\theta,\phi)] \Big\} \,.\, (-\mathbf{u}_r) \exp(+jkr\,\mathbf{u}_i \,.\, \mathbf{u}_r)\, r^2 \sin\theta \,\mathrm{d}\theta \mathrm{d}\phi \quad . \tag{4.40}$$

This complicated looking equation can be explained fairly simply in physical terms. Referring to Fig. 4.6, it is the interaction (roughly speaking, by means of the scalar product of the fields) of the incoming plane wave and the outward-travelling spherical wave. As kr is made larger and larger by letting $r \to \infty$, the spherical wave becomes more plane, and hence interacts in the single direction

$\mathbf{u}_r = \mathbf{u}_i$ more effectively with the incoming plane wave. In all other directions large phase variations cause the interaction to be self-destructive. Alternatively one can think of the spherical wave as an angular spectrum of plane waves, and the component plane wave travelling in the direction opposite to that of the incoming plane wave is the only one with which it will interact. Either way, the import of this physical picture is that the integral of equation (4.40) will be dominated by the value of $\mathbf{e}(\theta,\phi)$ in the direction of the incoming wave, namely be $\mathbf{e}(\mathbf{u}_i)$.

As the above physical argument suggests, the integral of equation (4.40) can be evaluated for very large kr by the method of stationary phase. Such a solution will be asymptotically exact as $kr \to \infty$, which is also the condition under which the far field of equation (4.38) is exact. There are two stationary-phase points, namely $\mathbf{u}_r = \pm\, \mathbf{u}_i$, but the negative sign reduces the integral to zero. Applying the algorithm for the stationary-phase evaluation of double integrals of equation (3.59), equation (4.40) becomes

$$\iint_{A_r} (\mathbf{E}_1 \times \mathbf{H}_2 - \mathbf{E}_2 \times \mathbf{H}_1) \,.\, \mathbf{n}\, \mathrm{d}A$$

$$= \frac{j\lambda^2}{\pi Z}\, \mathbf{e}_i \,.\, \mathbf{e}(\mathbf{u}_i) \quad . \tag{4.41}$$

Combining equations (4.37) and (4.41) and equating their sum to zero, by virtue of the Lorentz integral (4.33), the ratio of the signal amplitudes on reception and on transmission is

$$\frac{b}{a} = \frac{\lambda^2}{j4\pi Z P_o}\, \mathbf{e}_i \,.\, \mathbf{e}(\mathbf{u}_i) \tag{4.42}$$

which is an exact result.

Comment

The result of equation (4.42) is not affected by re-radiation from the antenna when illuminated by the incident plane wave, as the following argument shows. The re-radiation will have a far field of the same form as equation (4.38) for the directly radiated field, but with the vector pattern function $\mathbf{e}(\theta,\phi)$ replaced by $\mathbf{e}_s(\theta,\phi)$, which will in general be different from $\mathbf{e}(\theta,\phi)$. Hence in assessing the possible contribution of the re-radiated (that is, scattered) field to the Lorentz integral of equation (4.33) we retain equation (4.38) as the subscript 1 fields and take the subscript 2 fields to be

$$\mathbf{E}_2(r,\theta,\phi) = \frac{\exp(-jkr)}{kr}\, \mathbf{e}_s(\theta,\phi)$$

and

$$\mathbf{H}_2(r,\theta,\phi) = \frac{1}{Z}\, \mathbf{u}_r \times \mathbf{E}_2(r,\theta,\phi) \tag{4.43}$$

with

$$\mathbf{u}_r \cdot \mathbf{E}_2(r,\theta,\phi) = 0 \quad .$$

Then over the sphere of radius r, as $kr \to \infty$,

$$\iint_{A_r} (\mathbf{E}_1 \times \mathbf{H}_2 - \mathbf{E}_2 \times \mathbf{H}_1) \cdot \mathbf{n}\, dA$$

$$= \frac{\exp(-j2kr)}{Z(kr)^2} \int_{-\pi/2}^{\pi/2} \int_{-\pi}^{\pi} \Big\{ \mathbf{e}(\theta,\phi) \times [\mathbf{u}_r \times \mathbf{e}_s(\theta,\phi)]$$

$$- \mathbf{e}_s(\theta,\phi) \times [\mathbf{u}_r \times \mathbf{e}(\theta,\phi)] \Big\}\, r^2 \sin\theta \, d\theta d\phi \tag{4.44}$$

which is zero, since equation (2.13) shows that the integrand is identically zero. Hence re-radiation from the antenna in the form of scattering of the incident plane wave does not affect the reciprocity result of equation (4.42).

Exercise

Check the validity of equation (4.41).

4.2.2 Relation between effective receiving area and gain function

With the antenna reciprocity theorem of equation (4.42) we are now in a position to prove that equation (4.30) gives the relation between gain function and effective receiving area for an antenna, and to say precisely what is meant by polarization match.

According to equation (4.42), the power delivered to the antenna when it is illuminated by a plane wave $\mathbf{e}_i$ arriving from the direction $\mathbf{u}_i$ is

$$P_{\text{rec}} = \frac{\lambda^4 P_o}{16\pi^2 Z^2 P_o^{\,2}}\, |\mathbf{e}_i \cdot \mathbf{e}(\mathbf{u}_i)|^2 \tag{4.45}$$

which will be a maximum when the polarization of the incident plane wave $\mathbf{e}_i$ is identical to that of the vector pattern function $\mathbf{e}(\mathbf{u}_i)$ on transmission in the direction $\mathbf{u}_i$. When this occurs

$$|\mathbf{e}_i \cdot \mathbf{e}(\mathbf{u}_i)|^2 = |e_i|^2 \, |e(\mathbf{u}_i)|^2 \quad , \tag{4.46}$$

which is the condition known as **polarization match**. Hence dividing the maximum received power by the power density of the incident wave gives (by definition) the effective receiving area of the antenna as

$$A(\theta,\phi) = \frac{\lambda^4}{8\pi^2 Z P_o} \, |e(\theta,\phi)|^2 \quad . \tag{4.47}$$

This yields, on substituting the gain function for the antenna of equation (4.6)

$$A(\theta,\phi) = \frac{\lambda^2}{4\pi} \, G(\theta,\phi) \quad , \tag{4.48}$$

which is the formula already quoted as equation (4.30), and applies to any antenna.

In terms of the angular spectrum $F_x(\alpha,\beta)$ of the field radiated from an x-polarized aperture field, the effective receiving area in a direction (θ,ϕ) pointing into the half-space $z \geqslant 0$ is, from equations (4.10) and (4.48),

$$A(\theta,\phi) = \frac{\lambda^4}{2ZP_o} \, |F_x(\sin\theta\cos\phi, \sin\theta\sin\phi)|^2 (1 - \sin^2\theta\sin^2\phi) \quad . \tag{4.49}$$

4.3 GENERAL ANTENNA COUPLING FORMULA

One of the merits of the antenna reciprocity theorem of equation (4.42) is that it preserves information about the phase of the signal arriving at the receiver terminals. If more than one plane wave is incident on the antenna, then the complex signal received is given by the phasor sum of appropriate terms of the form of equation (4.42). In particular, if an angular spectrum of plane waves is incident on an antenna then the complex signal received is given by an integral, over all directions, of the scalar product of the incident angular spectrum and the vector pattern function of the antenna on transmission. We will apply this notion to determine the complex signal induced at the terminals of a receiving antenna when illuminated by a transmitting antenna an arbitrary distance away.

In order to keep the analysis fairly simple, suppose that a transmitting antenna T and receiving antenna R face each other on the squint across the semi-infinte slab shown in Fig. 4.8, and that both antennas are x-polarized with their respective x-directions into the paper. Let the (unprimed) coordinate system of the transmitter be based on O in its aperture, and the (primed) co-ordinate system of the receiver be based on O′ in its aperture; and let the vector distance between O and O′ be $\mathbf{r}_o$, and the direction OO′ have the direction cosines $(\alpha_o,\beta_o,\gamma_o)$ with respect to O.

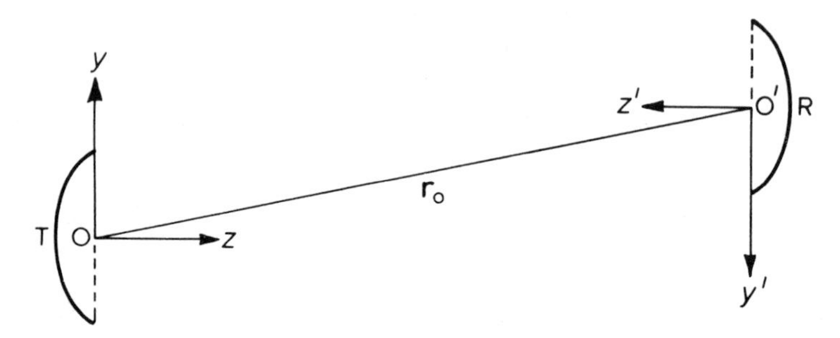

Fig. 4.8 – Transmitting antenna T and receiving antenna R facing each other on the squint across the semi-infinite slab between the planes $z = 0$ and $z' = 0$.

Assume that the transmitted angular spectrum of the transmitting antenna, with unit power delivered and radiated, is $F_x{}^T(\alpha,\beta)$. A representative elemental plane wave radiated by the transmitter will therefore have a vector electric field (see equation 3.11)

$$\mathbf{e} = F_x{}^{\mathrm{T}}(\alpha,\beta)\left[\mathbf{u}_x - \mathbf{u}_z\,\frac{\alpha}{\gamma}\right] d\alpha d\beta \tag{4.50}$$

referred to the point O as phase reference. Changing the phase-reference point from O to O′,

$$\mathbf{e} = F_x{}^{\mathrm{T}}(\alpha,\beta)\left[\mathbf{u}_x - \mathbf{u}_z\,\frac{\alpha}{\gamma}\right] \exp\{-jkr_o(\alpha_o\alpha + \beta_o\beta + \gamma_o\gamma)\}\, d\alpha d\beta \;. \tag{4.51}$$

Then, in the coordinate system of the receiving antenna, using the obvious transformations

$$\begin{array}{lcl} \mathbf{u}_x \longrightarrow \mathbf{u}_{x'} & & \alpha \longrightarrow -\alpha' \\ \mathbf{u}_y \longrightarrow -\mathbf{u}_{y'} & \text{and} & \beta \longrightarrow \beta' \\ \mathbf{u}_z \longrightarrow -\mathbf{u}_{z'} & & \gamma \longrightarrow \gamma' \end{array} \tag{4.52}$$

the elemental plane wave incident on the receiving antenna becomes

$$\mathbf{e} = -F_x{}^{\mathrm{T}}(-\alpha',\beta')\left[\mathbf{u}_{x'} - \mathbf{u}_{z'}\,\frac{\alpha'}{\gamma'}\right]$$
$$\exp\{-jkr_o(-\alpha_o\alpha' + \beta_o\beta' + \gamma_o\gamma')\}\, d\alpha' d\beta' \;. \tag{4.53}$$

If the angular spectrum of the receiving antenna when fed *and* radiating unit power is $F_x^R(\alpha',\beta')$, its vector pattern function on transmission with the field expressed in rectangular components (compare equation (3.61) with equation (4.1)), is

$$\mathbf{e}(\theta',\phi') = j2\pi\,[\mathbf{u}_{x'}\gamma' - \mathbf{u}_{z'}\alpha']\,F_x^{\mathrm{R}}(\alpha',\beta') \;. \tag{4.54}$$

The scalar product of the elemental transmitted wave of equation (4.53) and this vector pattern function is

$$\mathbf{e}\,.\,\mathbf{e}\,(\theta',\phi') = -j2\pi\,F_x^{\mathrm{T}}(-\alpha',\beta')\;F_x^{\mathrm{R}}(\alpha',\beta')\;\frac{1-\beta'^2}{\gamma'}$$

$$\exp\{-jkr_o\,(-\alpha_o\alpha' + \beta_o\beta' + \gamma_o\gamma')\}\,\mathrm{d}\alpha'\mathrm{d}\beta' \tag{4.55}$$

which is clearly polarization matched for every direction. Then, using the antenna reciprocity theorem of equation (4.42) and integrating over all directions seen by the receiving antenna looking towards the semi-infinite slab, the complex received signal is given by

$$c = \frac{b}{a} = -\frac{\lambda^2}{2Z}\int_{-\infty}^{\infty}\int_{-\infty}^{\infty} F_x^{\mathrm{T}}(-\alpha',\beta')\,F_x^{\mathrm{R}}(\alpha',\beta')\;\frac{1-\beta'^2}{\gamma'}$$

$$\exp\{-jkr_o\,(-\alpha_o\alpha' + \beta_o\beta' + \gamma_o\gamma')\}\,\mathrm{d}\alpha'\mathrm{d}\beta' \;. \tag{4.56}$$

Or, equivalently, in terms of the direction cosines of the transmitting antenna's coordinate system,

$$c = \frac{\lambda^2}{2Z}\int_{-\infty}^{\infty}\int_{-\infty}^{\infty} F_x^{\mathrm{T}}(\alpha,\beta)\,F_x^{\mathrm{R}}(-\alpha,\beta)\;\frac{1-\beta^2}{\gamma}$$

$$\exp\{-jkr_o(\alpha_o\alpha + \beta_o\beta + \gamma_o\gamma)\}\,\mathrm{d}\alpha\mathrm{d}\beta \;. \tag{4.57}$$

Equation (4.57) (or equation (4.56)) is the general formula for the signal coupled from one antenna to another.

Since it was assumed in the course of the above analysis that the transmitter radiates unit power, and also that the signal amplitude a (see antenna reciprocity theorem proof) corresponds to unit power travelling down the receiver feeder, it follows that

$$|c|^2 = \frac{P_{\mathrm{R}}}{P_{\mathrm{T}}} \;, \tag{4.58}$$

the ratio of power received (P_{R}) to power transmitted (P_{T}), often referred to as the **free-space transmission loss**.

Far-Field Coupling

As $kr_o \to \infty$, equation (4.57) may be evaluated by the method of stationary phase (using the algorithm of equation (3.59)) to give the received signal asymptotically as

$$c = \frac{j\lambda^3}{2Zr_o} (1 - \beta_o^2) F_x{}^{T}(\alpha_o,\beta_o)\; F_x{}^{R}(-\alpha_o,\beta_o)\, e^{-jkr_o} \tag{4.59}$$

where α_o, β_o, γ_o are the direction cosines from the transmitter to the receiver, and r_o is the distance between them. The far-field transmission loss is therefore

$$|c|^2 = \frac{\lambda^6}{4Z^2 r_o{}^2} (1 - \beta_o{}^2)^2 |F_x{}^{T}(\alpha_o,\beta_o)|^2 |F_x{}^{R}(-\alpha_o,\beta_o)|^2 \; . \tag{4.60}$$

Substituting the gain function $G_T(\theta,\phi)$ of the transmitter (see equation (4.10) and the effective receiving area $A_R(\theta,\phi)$ of the receiver (see equation (4.49)), both for unit power, the far-field transmission loss becomes

$$|c|^2 = \frac{G(\theta_o,\phi_o) A_R(\theta_o,\pi\text{-}\phi_o)}{4\pi r_o{}^2} \tag{4.61}$$

in which (θ_o,ϕ_o) are the polar and azimuth angles of the direction of the receiver from the transmitter. Equation (4.61) is the Friis far-field coupling formula of equation (4.31).

The geometry of Fig. 4.8 differs slightly from that originally used by Brown (1958a) in developing the general antenna coupling equation. The motivation for this original work was to study the effect of an arbitrary receiving antenna pattern on the measurement of the field radiated by a transmitting antenna. The appropriate geometry for this requires the receiving antenna always to be pointing towards the transmitter, which complicates both the coordinate transformations from O to O′ and the question of polarization match. The rather awkward-looking geometry of Fig. 4.8 was chosen here to skirt these complications without materially altering the conclusions.

Example

Consider the coupling between two antennas facing each other along their axes, separated by a distance $\mathbf{r}_o$, both of whose aperture fields on transmission are Gaussian in shape, as sketched in Fig. 4.9. If the aperture field of an antenna is

$$E_{ax}(x,y) = \frac{2}{W}\left(\frac{Z}{\pi}\right)^{1/2} \exp\left(-\frac{x^2+y^2}{W^2}\right) \tag{4.62}$$

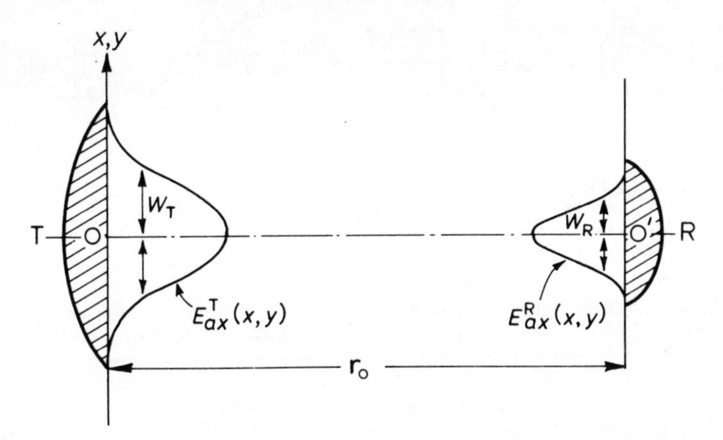

Fig. 4.9 – On-axis coupling between T and R whose aperture fields are Gaussian shaped.

it transmits unit power and has an angular spectrum

$$F_x(\alpha,\beta) = \frac{2W(\pi Z)^{1/2}}{\lambda^2} \exp\left(-\frac{\pi^2 W^2(\alpha^2+\beta^2)}{\lambda^2}\right) . \tag{4.63}$$

Assume that equation (4.62) applies to the transmitting antenna with $W = W_T$, and also to the receiving antenna with $W = W_R$. Then according to equation (4.57) the coupling between the antennas is

$$c = \frac{2\pi W_T W_R}{\lambda^2} \int_{-\infty}^{\infty}\int_{-\infty}^{\infty} \frac{1-\beta^2}{\gamma} \exp\left\{-\frac{\pi^2}{\lambda^2}(W_T^2 + W_R^2)(\alpha^2+\beta^2)\right\}$$

$$\exp(-jkr_o\gamma)\mathrm{d}\alpha\mathrm{d}\beta \quad . \tag{4.64}$$

If both T and R are narrow-beam antennas, such that both W_R and W_T measure several wavelengths, we may make the approximations that

$$\gamma \approx 1 - \tfrac{1}{2}(\alpha^2+\beta^2) \text{ and } \frac{1-\beta^2}{\gamma} \approx 1 \quad . \tag{4.65}$$

Then

$$c = \frac{2\pi W_T W_R}{\lambda^2} \exp(-jkr_o)$$

$$\int_{-\infty}^{\infty}\int_{-\infty}^{\infty} \exp\left\{-\left[\frac{\pi^2(W_T^2 + W_R^2)}{\lambda^2} - j\frac{\pi r_o}{\lambda}\right](\alpha^2 + \beta^2)\right\} d\alpha d\beta \tag{4.66}$$

$$= \frac{2\pi W_T W_R \exp(-jkr_o)}{\pi(W_T^2 + W_R^2) - j\lambda r_o} , \tag{4.67}$$

using the standard integral of equation (3.58).

The ratio of power received to power transmitted is therefore

$$|c|^2 = \frac{4\pi^2 W_T^2 W_R^2}{\pi^2(W_R^2 + W_T^2) + \lambda^2 r_o^2} , \tag{4.68}$$

a result obtained by T.S. Chu (1971) using another method.

In the limit as $r_o \to \infty$, we obtain the far-field result

$$|c|^2_{r_o \to \infty} = \frac{4\pi^2 W_T^2 W_R^2}{\lambda^2 r_o^2} \tag{4.69}$$

which agrees with Friis's formula of equation (4.31) if equations (4.10), (4.49) and (4.63) are used to find the gain function of T and the effective receiving area of R.

At the other extreme, as $r_o \to 0$,

$$|c|^2_{r_o \to 0} = \frac{4W_T^2 W_R^2}{(W_R^2 + W_T^2)^2} \tag{4.70}$$

which becomes equal to unity when $W_T = W_R$, which is correct physically.

The breakpoint between these two extremes occurs at a separation

$$r_o = \frac{\pi(W_R^2 + W_T^2)}{\lambda} . \tag{4.71}$$

Exercise

If the complete radiation pattern of the transmitter is defined by the two angular spectra $F_x{}^T(\alpha,\beta)$, $F_y{}^T(\alpha,\beta)$ and that of the receiver by $F_x{}^R(\alpha,\beta)$, $F_y{}^R(\alpha,\beta)$, show that the coupling between them for the geometry of Fig. 4.8 is

$$c = \frac{\lambda^2}{2Z} \int_{-\infty}^{\infty} \int_{-\infty}^{\infty} \left[\frac{1-\beta^2}{\gamma} F_x^{\,\mathrm{T}}(\alpha,\beta)\, F_x^{\,\mathrm{R}}(-\alpha,\beta) - \frac{\alpha\beta}{\gamma} F_x^{\,\mathrm{T}}(\alpha,\beta)\, F_y^{\,\mathrm{R}}(\alpha,\beta) \right.$$

$$\left. + \frac{\alpha\beta}{\gamma} F_y^{\,\mathrm{T}}(\alpha,\beta)\, F_x^{\,\mathrm{R}}(-\alpha,\beta) - \frac{1-\alpha^2}{\gamma} F_y^{\,\mathrm{T}}(\alpha,\beta)\, F_y^{\,\mathrm{R}}(-\alpha,\beta) \right]$$

$$\exp\{-jkr_o\,(\alpha_o\alpha + \beta_o\beta + \gamma_o\gamma)\}\, \mathrm{d}\alpha\mathrm{d}\beta \quad . \qquad (4.72)$$

4.3.1 Near-field antenna measurements

The antenna coupling formula can be applied directly to measurements made over a plane in the Fresnel region using a probing antenna R, to yield the angular spectrum (and hence the far-field pattern) of antenna T (see Fig. 4.10). This is

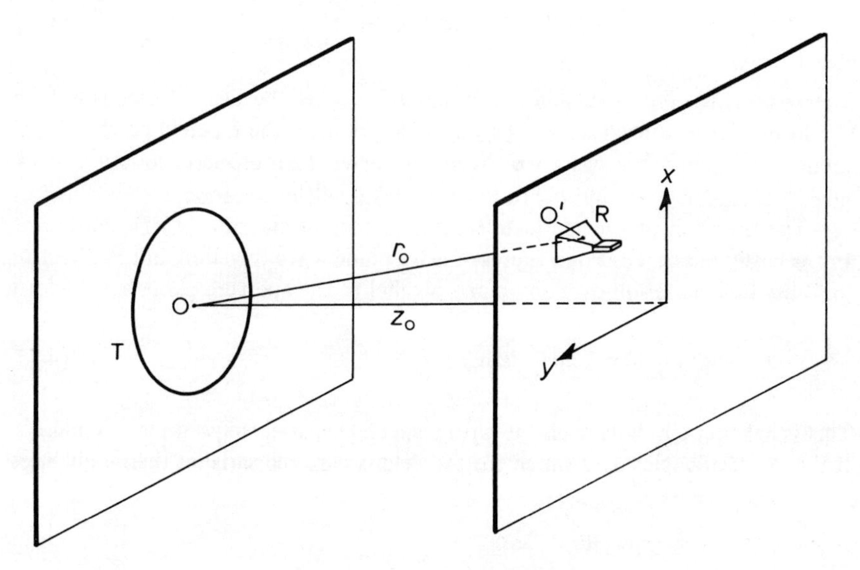

Fig. 4.10 – The field of transmitting antenna T being probed over a plane in the Fresnel region by receiving antenna R.

the simplest form taken by the so-called method of near-field antenna measurements, and is particularly suitable when T is a highly directive antenna.

Inserting the direction cosines

$$\alpha_o = \frac{x}{r_o}\ , \quad \beta_o = \frac{y}{r_o}\ , \quad \gamma_o = \frac{z_o}{r_o}$$

into equation (4.57), the coupled signal $c(x,y)$ when both T and R are x-polarized and R is at the point (x,y) in the plane, is

$$c(x,y) = \frac{\lambda^2}{2Z} \int_{-\infty}^{\infty} \int_{-\infty}^{\infty} F_x^{\,T}(\alpha,\beta)\; F_x^{\,R}(-\alpha,\beta) \frac{1-\beta^2}{\gamma} \exp\,(-jk\gamma z_o)$$

$$\exp\,\{-jk\,(\alpha x + \beta y)\}\; d\alpha d\beta \quad . \tag{4.73}$$

This is a two-dimensional Fourier transform which can be inverted immediately to give

$$F_x^{\,T}(\alpha,\beta)\; F_x^{\,R}(-\alpha,\beta) \frac{1-\beta^2}{\gamma} \exp\,(-jk\gamma z_o) = \frac{1}{2Z} \int_{-\infty}^{\infty} \int_{-\infty}^{\infty} c(x,y)$$

$$\exp\,\{+jk\,(\alpha x + \beta y)\}\; dx dy \quad . \tag{4.74}$$

The right-hand side is calculated, in practice, using the fast Fourier transform (FFT) on the measured signal $c\,(x,y)$. Then, provided the far-field pattern of the probe R is known and has no zeros in the forward hemisphere, equation (4.74) gives the angular spectrum $F_x^{\,T}(\alpha,\beta)$ of the transmitting antenna.

The question of the adequate spacing of the samples of $c\,(x,y)$ is important, and is easily answered. For a representative plane wave travelling in the direction (α,β) the field variation over any plane parallel to the aperture plane has the form

$$\exp\,(-jk\alpha x)\exp\,(-jk\beta y) \quad . \tag{4.75}$$

This is harmonic in both x and y, with respective spatial frequencies α/λ and β/λ. If the range of angles over which the far-field is required satisfies the inequalities

$$|\alpha| \leqslant \alpha_{max},\; |\beta| \leqslant \beta_{max} \quad , \tag{4.76}$$

then Nyquist's sampling criterion requires that the spacing between samples in the x and y directions should be

$$\Delta x = \frac{\lambda}{2\alpha_{max}} \quad \text{and} \quad \Delta y = \frac{\lambda}{2\beta_{max}} \quad . \tag{4.77}$$

The minimum required spacing, corresponding to coverage of the entire forward hemisphere, is clearly $\lambda/2$.

If the transmitting and receiving-probe patterns are not purely x-polarized the more general result of equation (4.72) applies. This shows that, even if the receiving probe is x-polarized and the transmitter is not, the y-polarized part of the transmitted field will contribute to the coupled signal. But it should be observed that the cross-coupled signals will be small for narrow-beam transmitting antennas.

Probing surfaces other than planes are often used in near-field measurements. The same general technique can be used to find the coupled signal, but considerable care must be taken in defining the relationship between the coordinates, direction cosines and unit vectors for the transmitter and receiver in such cases. A further complication is that an x-polarized transmitted plane wave will present itself to the receiver as a combination of x-polarized and y-polarized plane waves, except in the special case of a probing surface that is plane and parallel to the aperture plane (Fig. 4.10).

Exercise

Investigate the signal coupled from a transmitter to a probing receiver, both x-polarized, when the probing surface is (a) cylindrical and (b) spherical.

4.3.2 Aperture-field form of the antenna coupling formula

It is sometimes useful to have the antenna coupling formula of equation (4.56), or its equivalent equation (4.57), in terms of fields rather angular spectra. This is achieved by replacing the transmitter and receiver angular spectra. $F_x{}^{\mathrm{T}}(\alpha,\beta)$ and $F_x{}^{\mathrm{R}}(\alpha,\beta)$, by the Fourier transforms of the corresponding radiated fields over the aperture plane of the receiver. The appropriate geometry is that of Fig. 4.11.

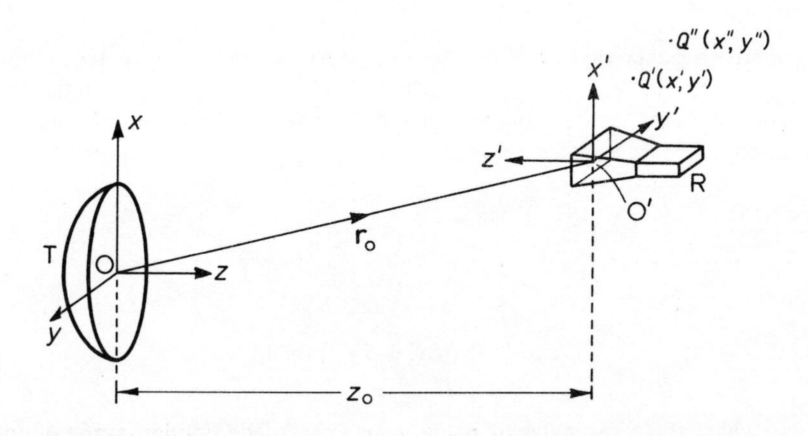

Fig. 4.11 – Geometry for the aperture-field form of antenna coupling.

Consider a point Q″, whose coordinates are (x'', y'') in the receiver plane $x'O'y'$. The tangential electric field there due to the x-polarized transmitter at O is

$$E_x{}^{\mathrm{T}}(\mathrm{Q}'') = \int_{-\infty}^{\infty}\int_{-\infty}^{\infty} F_x{}^{\mathrm{T}}(\alpha,\beta)\exp\{-jk[\alpha(x_o + x'') + \beta(y_o - y'') + \gamma z_o]\}\, d\alpha d\beta \tag{4.78}$$

since in the transmitter's Cartesian coordinate system the coordinates of Q″ are $(x_o + x'',\ y_o - y'',\ z_o)$. In terms of the receiver's direction cosines (α', β') this equation becomes, by virtue of transformation (4.52).

$$E_x{}^{\mathrm{T}}(\mathrm{Q}'') = -\int_{-\infty}^{\infty}\int_{-\infty}^{\infty} F_x{}^{\mathrm{T}}(-\alpha',\beta')\exp\{-jk(-\alpha' x_o + \beta' y_o + \gamma' z_o)\} \exp\{-jk(-\alpha' x'' - \beta' y'')\}\, d\alpha' d\beta' \ . \tag{4.79}$$

The Fourier inverse of this equation is

$$F_x{}^{\mathrm{T}}(-\alpha',\beta')\exp\{-jk(-\alpha' x_o + \beta' y_o + \gamma' z_o)\} = -\frac{1}{\lambda^2}\int_{-\infty}^{\infty}\int_{-\infty}^{\infty} E_x{}^{\mathrm{T}}(x'',y'')\exp\{+jk(-\alpha' x'' - \beta' y'')\}\, dx'' dy'' \tag{4.80}$$

in which $E_x{}^{\mathrm{T}}(x'',y'')$ could equally well be written as $E_{x'}{}^{\mathrm{T}}(x'',y'')$.

The y'-component of the magnetic field at point Q′ in the receiver's plane, corresponding to when the x-polarized receiver is transmitting, is, from equation (3.85),

$$H_{ay'}{}^{\mathrm{R}}(\mathrm{Q}') = \frac{1}{Z}\int_{-\infty}^{\infty}\int_{-\infty}^{\infty} \frac{1-\beta'^2}{\gamma'} F_x{}^{\mathrm{R}}(\alpha',\beta') \exp\{-jk(\alpha' x' + \beta' y')\}\, d\alpha' d\beta' \tag{4.81}$$

in which the coordinates of point Q′ are (x', y'). The Fourier inverse of equation (4.81) is

$$F_x{}^{\mathrm{R}}(\alpha',\beta')\frac{1-\beta'^2}{\gamma'} = \frac{Z}{\lambda^2}\int_{-\infty}^{\infty}\int_{-\infty}^{\infty} H_{ay'}{}^{\mathrm{R}}(x',y')$$

$$\exp\{jk\,(\alpha' x' + \beta' y')\}\,\mathrm{d}x'\mathrm{d}y' \quad . \tag{4.82}$$

Substituting equations (4.80) and (4.82) into the antenna coupling formula (4.56) the signal coupled into the receiver is given by the six-fold integral

$$c = \frac{1}{2\lambda^2}\int_{-\infty}^{\infty} (6) \int_{-\infty}^{\infty} \mathrm{d}\alpha'\mathrm{d}\beta'\,\mathrm{d}x''\mathrm{d}y''\,\mathrm{d}x'\mathrm{d}y'\,E_{x'}{}^{\mathrm{T}}(x'',y'')$$

$$H_{ay'}{}^{\mathrm{R}}(x',y')\exp\{jk\,[\alpha'(x'-x'') + \beta'(y'-y'')]\} \quad . \tag{4.83}$$

Taking the α',β' integrations first, which is justified by the reasonable assumption that both $E_{x'}{}^{\mathrm{T}}$ and $H_{ay'}{}^{\mathrm{R}}$ are continuous, equation (4.83) becomes

$$c = \tfrac{1}{2}\int_{-\infty}^{\infty} (4) \int_{-\infty}^{\infty} \mathrm{d}x''\mathrm{d}y''\mathrm{d}x'\mathrm{d}y'\;E_{x'}{}^{\mathrm{T}}(x'',y'')$$

$$H_{ay'}{}^{\mathrm{R}}(x',y')\,\delta(x'-x'')\,\delta(y'-y'') \tag{4.84}$$

which gives the coupled signal as

$$c = \tfrac{1}{2}\int_{-\infty}^{\infty}\int_{-\infty}^{\infty} E_{x'}{}^{\mathrm{T}}(x',y')\,H_{ay'}{}^{\mathrm{R}}(x',y')\,\mathrm{d}x'\mathrm{d}y' \quad . \tag{4.85}$$

This is a useful alternative to the angular-spectrum forms for the antenna reciprocity theorem of equations (4.56) and (4.57).

Exercise

Deduce from equation (4.72) and the above analysis that the total signal coupled to the receiver, when T and R have both x- and y- polarized components of tangential electric field in their aperture fields, is

$$c = \tfrac{1}{2}\int_{-\infty}^{\infty}\int_{-\infty}^{\infty} [E_{x'}{}^{\mathrm{T}}H_{ay'}{}^{\mathrm{R}} - E_{y'}{}^{\mathrm{T}}H_{ax'}{}^{\mathrm{R}}$$

$$- H_{y'}{}^{\mathrm{T}}E_{ax'}{}^{\mathrm{R}} + H_{x'}{}^{\mathrm{T}}E_{ay'}{}^{\mathrm{R}}]\,\mathrm{d}x'\mathrm{d}y' \quad . \tag{4.86}$$

Show that this coupling formula can be expressed more succinctly as

$$c = \tfrac{1}{2} \int_{a'} (\mathbf{E}^{\mathrm{T}} \times \mathbf{H}^{\mathrm{R}} - \mathbf{E}^{\mathrm{R}} \times \mathbf{H}^{\mathrm{T}}) \,.\, \mathbf{n}' \,\mathrm{d}a' \quad , \tag{4.87}$$

where $\mathbf{E}^{\mathrm{T}}, \mathbf{H}^{\mathrm{T}}$ are the fields in the aperture plane of the receiver due to the transmitter, $\mathbf{E}^{\mathrm{R}}, \mathbf{H}^{\mathrm{R}}$ are the fields there due to the receiver when transmitting, $\mathbf{n}'$ is normal to the receiver aperture plane pointing towards the transmitter, and $\mathrm{d}a'$ is an element of area in the receiver aperture plane.

CHAPTER 5

Fresnel diffraction

It is useful to distinguish three regions in the field diffracted from a radiating aperture. These are the reactive, or evanescent-wave, region; the Fresnel region; and the far-field, or Fraunhofer, region. The three regions are indicated in Fig. 5.1.

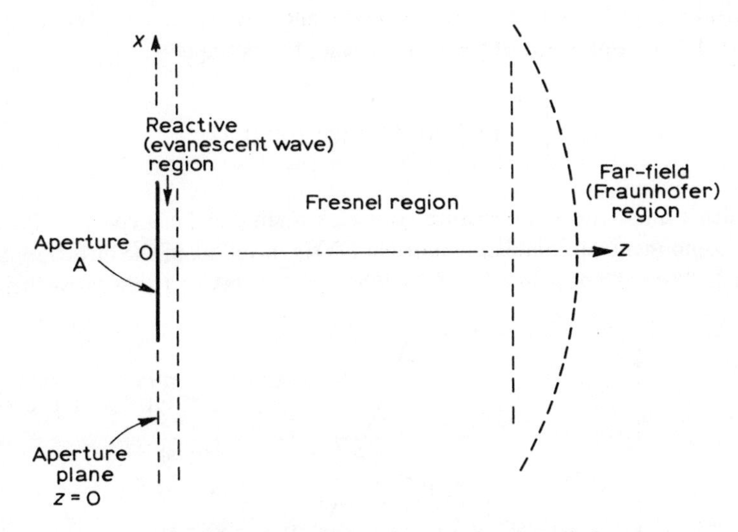

Fig. 5.1 – Three regions for the field diffracted from a radiating aperture.

Proceeding from the aperture plane $z = 0$ into the half-space $z \geqslant 0$, the first region encountered is the reactive region in which the evanescent waves as well as the propagating waves in the angular plane-wave spectrum contribute to the field. As demonstrated in sections 2.3 and 3.3 the reactive region is only a few wavelengths wide. Beyond this lies the region of most general interest, in which all the propagating waves in the angular spectrum contribute to the field. This is the Fresnel region which extends out to many wavelengths beyond the radiating aperture. Eventually the Fresnel region shades into the far-field region, where the field is essentially determined by a single plane-wave component of the angular spectrum. This single component of the spectrum was demonstrated in sections 2.5 and 3.4 to be that value of the spectrum in the direction of the field point.

Most of the discussion of Fresnel diffraction in this chapter will be for two-dimensional fields of the transverse-magnetic (TM) type, whose field components are E_x, E_z and H_y. The final section (5.5) will extend the discussion to three dimensions, in which the principles are the same but the corresponding formulas are more complicated.

An alternative division of the diffracted field† is into the two broad categories of near field and far field, with the former subdivided into reactive near-field and radiating near-field regions. In this scheme the radiating near-field region corresponds exactly to the Fresnel region in the following discussion.

5.1 FRESNEL'S DIFFRACTION FORMULA

The three field components anywhere in the half-space $z \geqslant 0$ for a two-dimensional TM field are given in terms of the angular plane-wave spectrum $F(s)$ by equation (2.30). The x-component of the electric field, for example,

$$E_x(x,z) = \int_{-\infty}^{\infty} F(s) \exp\{-jk\,(sx + cz)\}\, ds \tag{5.1}$$

in which s and c are the direction cosines, $s = \sin\theta$ and $c = \cos\theta$, of the plane-wave component of complex amplitude $F(s)ds$ travelling in the direction making an angle θ to the z-axis (see Fig. 5.2). When $|s| > 1$, taking the negative imaginary

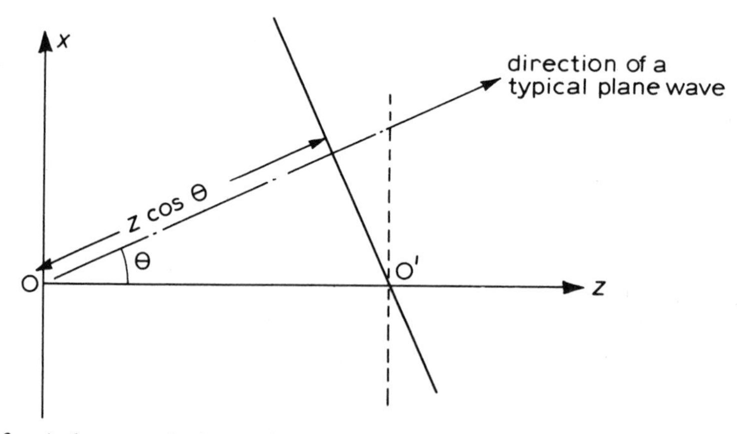

Fig. 5.2 – A plane wave in the angular spectrum referred to a new phase reference point O′.

†IEEE Standard Test Procedures for Antennas (149 – 1979).

root of $c = (1-s^2)^{1/2}$ means that the corresponding wave components in the angular spectrum are evanescent with increasing z. They will be ignored in the subsequent discussion.

Rearranging equation (5.1) slightly,

$$E_x(x,z) = \int_{-\infty}^{\infty} F(s) \exp(-jkcz) \exp(-jksx)\, ds$$

$$= \int_{-\infty}^{\infty} G(s) \exp(-jksx)\, ds \tag{5.2}$$

where

$$G(s) = F(s) \exp(-jkcz) \tag{5.3}$$

is the angular spectrum transformed from having the phase reference point O in the plane $z = 0$ to the point O′ for some arbitrary plane z. Then if we let

$$g(x) = E_x(x,z) \tag{5.4}$$

denote the x-component of the electric field over the plane z, equation (5.2) becomes

$$g(x) = \int_{-\infty}^{\infty} G(s) \exp(-jksx)\, ds \quad . \tag{5.5}$$

Thus $g(x)$ is the Fourier transform of $G(s)$, represented symbolically by

$$G(s) \longleftrightarrow g(x). \tag{5.6}$$

This last result is a generalization of the basic result derived in Chapter 2, that

$$F(s) \longleftrightarrow f(x) \tag{5.7}$$

which applies to the aperture field $f(x) = E_{ax}(x) = E_x(x,0)$ over the plane $z = 0$ and the angular plane-wave spectrum $F(s)$ referred to the point O in that plane.

An important consequence of the analysis up to this point is that if the tangential electric field $g(x)$ is known over any plane parallel to the aperture plane, then equations (5.5) and (5.3) can be inverted to yield

$$G(s) = \frac{1}{\lambda} \int_{-\infty}^{\infty} g(x) \exp(jksx)\, dx \tag{5.8}$$

and

$$F(s) = G(s) \exp(jkcz) \tag{5.9}$$

And since the orientation of the aperture plane is chosen arbitrarily we may claim that if the tangential field is known over *any* plane then the angular spectrum describing the propagating part of the field can be deduced.

What may be missing, depending on the distance from the aperture plane, is information about the evanescent part of the spectrum. But in many applications this information will be of no interest.

Another point to notice is that if the phase of the angular spectrum is of no consequence, for example when calculating the far-field *power* radiation pattern, then we have from (5.9) that

$$|F(s)| = |G(s)| \quad . \tag{5.10}$$

Returning now to equation (5.2) we look for an approximate evaluation of field components such as $E_x(x,z)$, applicable in the Fresnel region. This will be done in two ways: the first assuming that the angular spectrum is narrow, and the second from Huygens' principle. The result will be referred to as Fresnel's diffraction formula. Later on in the chapter the formula will be used to investigate diffraction of electromagnetic waves by a straight edge, and also how the field diffracted by an aperture changes as the wave propagates through the Fresnel region into the far field.

5.1.1 Fresnel's Diffraction Formula for Narrow Angular Spectra

It often happens that the angular spectrum $F(s)$ is of negligible amplitude outside a narrow range of angles. This is the case for many microwave antennas, for laser beams and diffraction by a straight edge, for example. Assuming that the relevent angular range is centred on $\theta = 0$, equation (5.3) can be written approximately as

$$G(s) = F(s) \exp \{-jkz(1 - \tfrac{1}{2}s^2)\} \quad . \tag{5.11}$$

This was achieved by expanding the cosine term $c = (1-s^2)^{1/2}$ using the binomial theorem, and discarding powers of s higher than the second, since s has been assumed small. Just how small s must be depends on the accuracy required. But it is worth noting that for $s = \frac{1}{6}$, for which θ is getting on for 10°, the approximate formula for c is in error by only one part in 10^4.

Substituting (5.11) into (5.2), the x component of the electric field is approximately

$$E_x(x,z) = \exp(-jkz) \int_{-\infty}^{\infty} F(s) \exp(j\tfrac{1}{2}kzs^2) \exp(-jkxs) \, ds. \tag{5.12}$$

This is the Fourier transform of the product of two functions of s, which themselves have the Fourier transforms

$$F(s) \longleftrightarrow f(x) \text{ (the aperture field)} \tag{5.13}$$

and

$$\exp\left(j\tfrac{1}{2}kzs^2\right) \longleftrightarrow (j\lambda/z)^{1/2} \exp\left(-j\tfrac{1}{2}kx^2/z\right) \; . \tag{5.14}$$

It follows from the convolution theorem of Fourier transforms that

$$g(x) = E_x(x,z) = (j/\lambda z)^{1/2} \exp(-jkz) \int_{-\infty}^{\infty} f(\xi) \exp\left\{-j\,\frac{k(x-\xi)^2}{2z}\right\} \mathrm{d}\xi \tag{5.15}$$

which is Fresnel's diffraction formula.

Exercise
Use the standard integral of equation (2.96) to confirm the validity of (5.14).

Exercise
Starting from equation (2.42) as defining the Fourier transform, deduce the precise form of the convolution theorem, and hence derive equation (5.15).

The only approximation made in deriving the above result is that the angular spectrum is negligible outside a narrow range of angles, say ± 10°. It will be shown later that this corresponds to the more fundamental approximation to the wave equation known as the **parabolic approximation.** But first it is instructive to obtain the same result by the more traditional method.

5.1.2 Fresnel's Diffraction Formula Derived from Huygens' Principle

Consider the same 2-dimensional TM case in which the aperture field $f(x) = E_{ax}(x)$ is defined over the plane $z = 0$. In the geometry of Fig. 5.3, a typical Huygens source can be taken as lying in the aperture plane at the point $x = \xi$ and of width $\mathrm{d}\xi$. The cylindrical wave emanating from this wavefront element is given by equation (2.83) as

$$\mathrm{d}E_\theta = Z\,\mathrm{d}H_y = f(\xi)\,\mathrm{d}\xi \sqrt{\frac{j}{\lambda r}}\,\mathrm{e}^{-jkr} \; . \tag{5.16}$$

Since $r = (z^2 + (x-\xi)^2)^{1/2}$, provided $\max |x-\xi| \ll z$, suitable approximations for r in equation (5.16) are:

$r \approx z\,[1 + (x-\xi)^2/(2z)]$ in the phase term,
$r \approx z$ in the amplitude term.

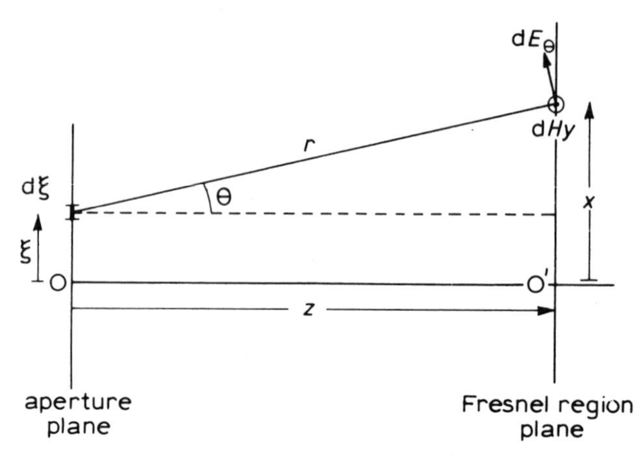

Fig. 5.3 – Field in the Fresnel region from a Huygens element in the aperture plane.

Similarly the inclination of the elemental electric-field vector dE_θ to the vertical may be ignored, and the x-component of the electric field in the Fresnel-region plane z can be written approximately as

$$g(x) = E_x(x,z) = \sqrt{\frac{j}{\lambda z}}\ \exp(-jkz) \int_{-\infty}^{\infty} f(\xi) \exp\{-jk(x-\xi)^2/(2z)\}\, d\xi \tag{5.17}$$

To this degree of approximation the only other significant field component is

$$H_y(x,z) = \frac{g(x)}{Z} \quad . \tag{5.18}$$

5.1.3 The Parabolic Equation Approximation

It is useful to show that Fresnel's diffraction formula is a solution of the parabolic equation (p.e.) approximation to the wave equation. This comes about because the physical assumption underlying the derivation of Fresnel's diffraction formula is that the fields are propagating essentially in a single direction. Assuming that the principal direction of propagation is the z-direction, we may write any one of the field components in the form (appropriate for a two-dimensional problem)

$$f(x,z) = u(x,z) \exp(-jkz) \tag{5.19}$$

in which $u(x,z)$ is to be seen as a slowly varying function of z, in contrast to $\exp(-jkz)$ which is rapidly varying.

All field components must satisfy the wave equation, which is (see Exercise B.2.5 (a)), for a source-free region and monochromatic fields,

$$\nabla^2 \mathbf{E} + k^2 \mathbf{E} = \mathbf{0} \ . \tag{5.20}$$

In the present instance

$$\frac{\partial^2 f}{\partial x^2} + \frac{\partial^2 f}{\partial z^2} + k^2 f = 0 \ , \tag{5.21}$$

which is a partial differential equation of elliptic type.

Now substituting (5.19) into (5.21)

$$\frac{\partial^2 u}{\partial x^2} + \frac{\partial^2 u}{\partial z^2} - j2k \frac{\partial u}{\partial z} = 0 \ . \tag{5.22}$$

Since u is a slowly varying function of z, we may take it that

$$\left| \frac{\partial^2 u}{\partial z^2} \right| \ll 2k \left| \frac{\partial u}{\partial z} \right| \tag{5.23}$$

which implies that u changes by a negligible amount in a distance $(2k)^{-1} = \lambda/(4\pi)$. Hence we have, approximately, that

$$\frac{\partial^2 u}{\partial x^2} - j2k \frac{\partial u}{\partial z} = 0 \ . \tag{5.24}$$

This is the parabolic equation approximation, so named because the approximation has yielded a partial differential equation of parabolic type.

It is a straightforward matter to show that Fresnel's diffraction formula satisfies the p.e. approximation. We may therefore reverse the argument and say that the p.e. approximation applies to fields whose angular spectra are narrowly confined to a single direction, the range of θ over which the spectrum is significant being no more than 20°.

5.2 DIFFRACTION BY A CONDUCTING HALF PLANE

Consider the two-dimensional problem, illustrated in Fig. 5.4, of an x-polarized plane wave of amplitude E_0 incident normally on a thin, perfectly conducting half plane. This classical problem in optics was first solved by Fresnel, and led to the establishment of the wave theory of light. Its solution has been used extensively by engineers as a model for radio propagation over such obstacles as mountain ridges and the edges of buildings. More recently it has been used in the application of the geometrical theory of diffraction, where it is the simplest of what are termed the 'canonical problems'.

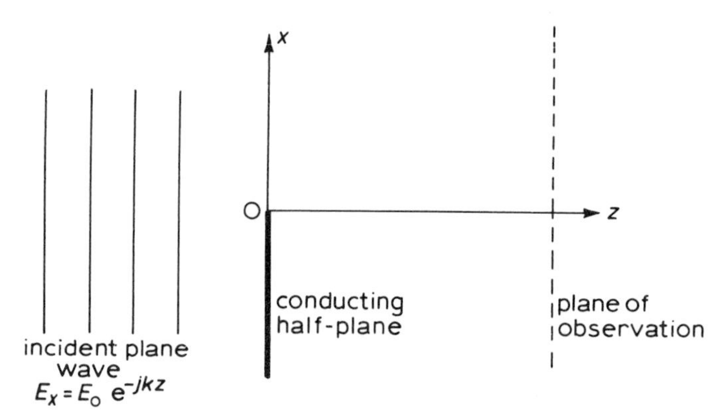

Fig. 5.4 – Plane wave incident normally on a conducting half plane.

Taking the aperture plane to be situated just to the right of the conducting half plane, at $z = 0^+$, the aperture field can be assumed as a first approximation to be

$$f(x) = E_{ax}(x) = E_o U(x) \tag{5.25}$$

where $U(x)$ is the Heaviside unit step. The corresponding angular plane-wave spectrum is

$$F(s) = E_o \left[\tfrac{1}{2}\delta(s) + j\frac{1}{2\pi s} \right] \tag{5.26}$$

A notable feature of this spectrum is its behaviour in the vicinity of $s = 0$. In addition to the delta function the second term, which is imaginary, is infinitely discontinuous at $s = 0$. An immediate consequence is that the far-field formula of equation (2.77) is not directly applicable, since it is based on the assumption that $F(s)$ is continuous. Or looking at it another way, the Rayleigh far-field criterion (see section 5.3.1) cannot be satisfied for an infinitely wide aperture. But since the angular spectrum is dominated by its behaviour in the region $s = 0$, the conditions for the validity of Fresnel's diffraction formula (5.15) are fulfilled.

Hence, substituting (5.25) into (5.15),

$$E_x(x,z) = E_o e^{-jkz} \sqrt{\frac{j}{\lambda z}} \int_0^\infty \exp\left\{ -j\frac{k(\xi - x)^2}{2z} \right\} d\xi \tag{5.27}$$

which can be evaluated in terms of tabulated Fresnel integrals (see below) by making the substitution

$$u = \sqrt{\frac{2}{\lambda z}}\,(\xi - x) \text{ with } \mathrm{d}\xi = \sqrt{\frac{\lambda z}{2}}\,\mathrm{d}u \tag{5.28}$$

so that

$$E_x(x,z) = E_o e^{-jkz}\sqrt{\frac{j}{2}}\int_{-x\sqrt{\frac{2}{\lambda z}}}^{\infty} \exp\left\{-j\,\frac{\pi u^2}{2}\right\}\mathrm{d}u$$

$$= E_o e^{-jkz}\sqrt{\frac{j}{2}}\left[\int_0^{\infty} \exp\left\{-j\,\frac{\pi u^2}{2}\right\}\mathrm{d}u + \int_0^{x\sqrt{\frac{2}{\lambda z}}} \exp\left\{-j\,\frac{\pi u^2}{2}\right\}\mathrm{d}u\right], \tag{5.29}$$

and

$$E_x(x,z) = E_o e^{-jkz}\sqrt{\frac{j}{2}}\left[\mathscr{F}(\infty) + \mathscr{F}\left(x\sqrt{\frac{2}{\lambda z}}\right)\right]. \tag{5.30}$$

5.2.1 The Fresnel Integral

In complex form, the Fresnel integral is

$$\mathscr{F}(v) = \int_0^v \exp\left\{-j\,\frac{\pi u^2}{2}\right\}\mathrm{d}u \tag{5.31}$$

$$= \mathscr{C}(v) - j\mathscr{S}(v) \tag{5.32}$$

where

$$\mathscr{C}(v) = \int_0^v \cos\left(\frac{\pi u^2}{2}\right)\mathrm{d}u \tag{5.33}$$

and

$$\mathscr{S}(v) = \int_0^v \sin\left(\frac{\pi u^2}{2}\right)\mathrm{d}u \tag{5.34}$$

are the widely tabulated Fresnel integrals in real form (see, for example, Abramowitz and Stegun (1964)). $\mathscr{F}(v)$ has the following properties:

$$\mathscr{F}(0) = 0 \tag{5.35}$$
$$\mathscr{F}(\nu) = -\mathscr{F}(-\nu) \tag{5.36}$$
$$\mathscr{F}(\infty) = -\mathscr{F}(-\infty) = \tfrac{1}{2}(1-j) \tag{5.37}$$

and asymptotically, as $\nu \to \infty$,

$$\mathscr{F}(\nu) \simeq \tfrac{1}{2}(1-j) - \frac{1}{j\pi\nu}\, e^{-j\frac{\pi\nu^2}{2}} \quad . \tag{5.38}$$

A plot of $\mathscr{S}(\nu)$ against $\mathscr{C}(\nu)$, with ν as a parameter, is shown in Fig. 5.5. It is known as **Cornu's spiral.** The radius vector OR marked on the diagram represents the complex conjugate of $\mathscr{F}(\nu)$ for a ν-value of about 1.3.

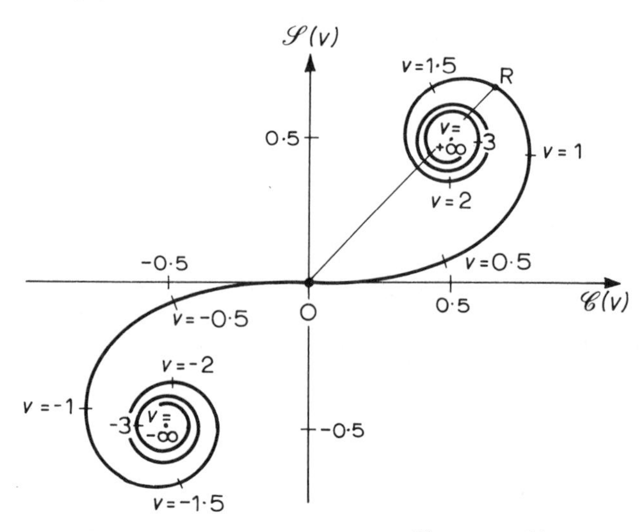

Fig. 5.5 – Cornu's spiral : a plot of $\mathscr{S}(\nu)$ against $\mathscr{C}(\nu)$.

Exercise
Show that the total length of Cornu's spiral, measured from the origin in either direction, is $|\nu|$, and that the slope at any ν is $\pi\nu^2/2$. Hence show that the radius of curvature of the spiral depends inversely on its length.

Exercise
Calculate the error made in taking

$$|\mathscr{F}(\tfrac{1}{2})| = \tfrac{1}{2} \qquad (0.69\%) \quad . \tag{5.39}$$

Exercise
Estimate the ν-values for which the magnitude of $\mathscr{F}(\nu)$ is (a) maximum and (b) minimum.

Exercise

Use the standard integral (2.67) to show that

$$\mathscr{F}(\infty) = \tfrac{1}{2}\,(1-j) \quad .$$

Equation (5.30) for the x-component of the electric field over the observation plane z, is conveniently written as

$$E_x(x,z) = E_o \mathrm{e}^{-jkz}\; W\left(x\sqrt{\frac{2}{\lambda z}}\right) \tag{5.40}$$

where

$$W\left(x\sqrt{\frac{2}{\lambda z}}\right) = \sqrt{\frac{j}{2}}\left[\mathscr{F}(\infty) + \mathscr{F}\left(x\sqrt{\frac{2}{\lambda z}}\right)\right] \tag{5.41}$$

is the (complex) ratio of the field at point (x,z) to its value in the absence of the conducting half-plane. Fig. 5.6 uses the Cornu spiral to represent $W(x\sqrt{(2/\lambda z)})$ as the radius vector OP.

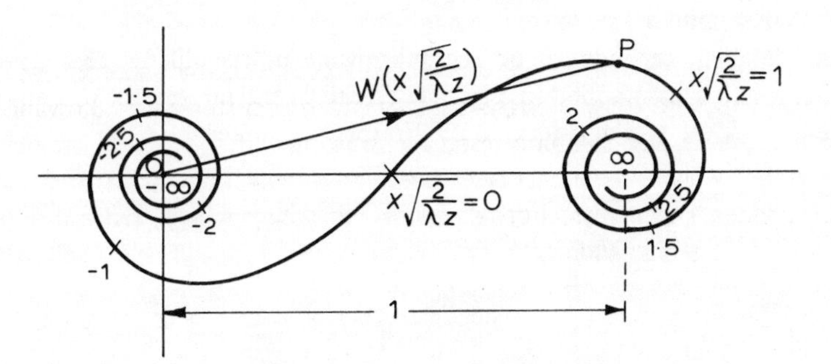

Fig. 5.6 – Ratio of diffracted field to incident field.

Supposing λ and z to be fixed, for large positive values of x

$$x\sqrt{\frac{2}{\lambda z}} \to \infty \text{ and } W\left(x\sqrt{\frac{2}{\lambda z}}\right) \to 1 \quad . \tag{5.42}$$

Hence the diffracted field approaches

$$E_x(x,z) = E_o \mathrm{e}^{-jkz} \text{ as } x \to \infty \quad , \tag{5.43}$$

which is identical to the incident plane wave. The same result is obtained for any positive x, no matter how small, in the goemetrical-optics limit $\lambda \to 0$. It is also obtained for finite λ and any positive x as $z \to 0$, where it is the aperture field.

On the z axis, which can be identified as the geometrical-optics shadow boundary,

$$x\sqrt{\frac{2}{\lambda z}} = 0 \text{ and } W = \tfrac{1}{2} \quad , \tag{5.44}$$

and hence

$$E_x(0,z) = \tfrac{1}{2}\, E_0 e^{-jkz} \quad , \tag{5.45}$$

which is just half the incident field.

For large negative values of x

$$x\sqrt{\frac{2}{\lambda z}} \to -\infty \;,\quad W\left(x\sqrt{\frac{2}{\lambda z}}\right) \to 0 \tag{5.46}$$

in such a way that the magnitude of the diffracted field goes monotonically to zero. Again, this limit is approached for any negative x, no matter how small, if either λ or z approaches zero.

The detailed dependence of the magnitude of the diffracted field on x, for fixed λ and z, is graphed in Fig. 5.7. The first, and largest, maximum occurs close to $x = \sqrt{(\lambda z)}$, which is termed the Fresnel parameter.

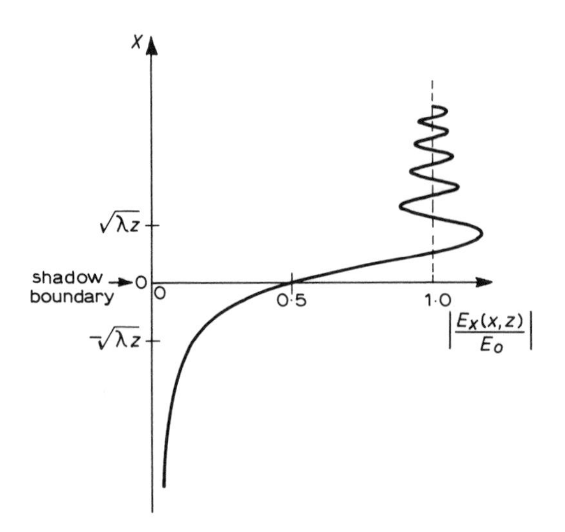

Fig. 5.7 – Graph of the magnitude of the diffracted field against distance x from the shadow boundary.

5.2.2 Diffracted Rays : The geometrical theory of diffraction

The solution we have obtained for the field in the shadow region has an interesting interpretation in terms of Keller's (1962) geometrical theory of diffraction. In this theory ordinary geometrical rays represent the plane wave incident on the conducting half plane, as indicated in Fig. 5.8. The theory applies in the limit of vanishingly small wavelength. Thus the field in the illuminated region is that due to the unobstructed incident rays, as we have already shown in equation (5.43).

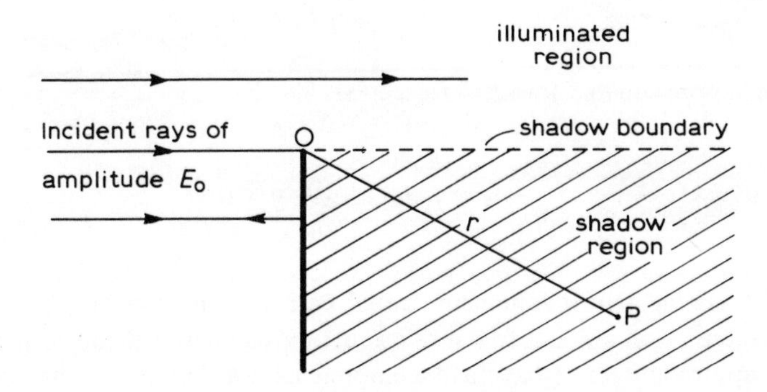

Fig. 5.8 – A ray diffracted into the shadow region.

In addition to this classical geometrical optics field, the geometrical theory of diffraction supposes that diffracted *rays* exist. One such is OP in Fig. 5.8. The ray incident on the edge O of the conducting half plane follows a straight-line path to a point such as P, thereby obeying an extension of Fermat's principle that the optical path should be a minimum. The scalar field at P, in the zero-wavelength limit, is assumed to be of the form

$$E_p = DE_o \frac{\exp(-jkr)}{\sqrt{r}} \tag{5.47}$$

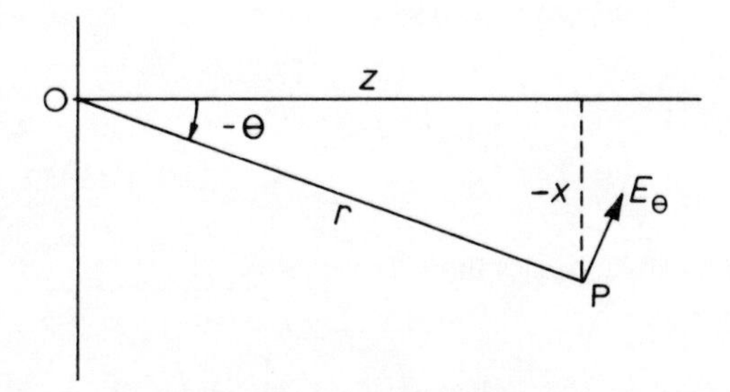

Fig. 5.9 – Geometry of the point P in the shadow region.

where D is termed the **diffraction coefficient.** The validity of this asymptotic form for the diffracted field, and the value of the diffraction coefficient D for a straight, conducting knife edge, will now be deduced from the previous analysis.

The field in the shadow region is given by equation (5.30) with x negative. In the limit as $\lambda \to 0$ the asymptotic form (5.38) for the complex Fresnel integral applies, and after some slight rearrangement

$$E_x(x,z) = \frac{jE_o}{2\pi \sin\theta} \sqrt{\frac{j\lambda}{z}} \exp\left\{-jk\left(z + \frac{x^2}{2z}\right)\right\} \cos\theta,\ x < 0 \ . \quad (5.48)$$

This is well approximated for small θ (see Fig. 5.9) by

$$E_x(r,\theta) = \frac{jE_o}{2\pi \sin\theta} \sqrt{\frac{j\lambda}{r}} \exp(-jkr) \cos\theta,\ \theta < 0 \ . \quad (5.49)$$

The remaining field components could be found in a similar, if somewhat more laborious, manner. But Maxwell's equation combined with the parabolic-equation approximation (that field components vary most rapidly in the z direction in a case such as this) give us the same information more elegantly. Thus (Appenix B, equation (B.2.29)) the divergence of electric field equation ($E_y = 0$ in this two-dimensional TM case)

$$\frac{\partial E_x}{\partial x} + \frac{\partial E_z}{\partial z} = 0 \quad (5.50)$$

yields the approximate relation

$$E_z = \frac{1}{jk} \frac{\partial E_x}{\partial x} \quad (5.51)$$

and from equation (5.48) with $\tan\theta = x/z$, we have

$$E_z = -\left[\frac{1}{jkx} + \frac{x}{z}\right] E_x \quad (5.52)$$

which, in the geometrical optics limit $k \to \infty$, is

$$E_z = -E_x \tan\theta = -\frac{jE_o}{2\pi \sin\theta} \sqrt{\frac{j\lambda}{z}} \exp\left\{-jk\left(z + \frac{x^2}{2z}\right)\right\} \sin\theta \ . \quad (5.53)$$

To find the remaining field component H_y, the first of Maxwell's curl equations (Appendix B, equation (B.2.31)) gives

$$H_y = \frac{1}{j\omega\mu}\left[\frac{\partial E_z}{\partial x} - \frac{\partial E_x}{\partial z}\right] . \tag{5.54}$$

The partial derivatives can be obtained from equations (5.48) and (5.53), and when substituted into (5.54) yield

$$H_y = \frac{E_x}{Z}\left[1 + \frac{x^2}{2z^2} - \frac{1}{jkz}\right] . \tag{5.55}$$

Then in the geometrical optics limit

$$H_y = \frac{E_x}{Z} \sec\theta . \tag{5.56}$$

Finally, collecting results, the field at the point P in the shadow region (Figs. 5.8 and 5.9) is given by

$$E_\theta(r,\theta) = ZH_y(r,\theta) = \sqrt{\frac{j\lambda}{r}} \; \frac{jE_o}{2\pi\sin\theta} \, e^{-jkr} , \tag{5.57}$$

which is a cylindrical wave emanating from the edge O of the diffracting half plane. This confirms the form of the diffracted ray given by equation (5.47) and gives it greater precision. The diffraction coefficient is

$$D = \sqrt{\frac{\lambda}{j}} \; \frac{1}{2\pi\sin\theta} . \tag{5.58}$$

The following conclusions can be drawn from the above analysis of the simplest canonical problem in the geometrical theory of diffraction, namely, diffraction by a thin conducting half plane of a normally incident plane wave.

(a) The field well into the illuminated region is identical to the incident field, and is given straightforwardly by classical geometrical optics.
(b) The field well into the dark region has the form of a cylindrical wave originating at the edge; its amplitude depends inversely on the sine of the angle between the diffracted ray and the shadow boundary.

(c) In the vicinity of the shadow boundary neither of the above prescriptions is valid, and the complete solution in terms of Fresnel's integrals must be used.

(d) It can be seen in retrospect that the field in the shadow region can be represented by the usual two-dimensional far-field formula

$$E_\theta(r,\theta) = ZH_y(r,\theta) = \sqrt{\frac{j\lambda}{r}}\, F(s)\, e^{-jkr} \quad , \tag{5.59}$$

with the angular spectrum (see equation (5.26))

$$F(s) = j\frac{E_o}{2\pi s} \quad , \quad s < 0 \quad , \tag{5.60}$$

substituted.

5.3 TRANSITION TO THE FAR FIELD

It was shown in section 2.5 that at sufficiently large distances from a radiating aperture of finite size the field acquires the simple far-field form of a spherical wave whose angular dependence is proportional to the angular spectrum. We will now use Fresnel's diffraction formula to determine the distance from the aperture at which this transition to the far field can reasonably be said to occur.

5.3.1 Rayleigh's far-field criterion

In two dimensions, with a wholly x-polarized aperture field $f(x) = E_x(x,0)$, Fresnel's diffraction formula (5.15) gives the field over an arbitrary plane in the Fresnel region as

$$g(x) = E_x(x,z) = \sqrt{\frac{j}{\lambda z}}\, \exp(-jkz) \int_{-\infty}^{\infty} f(x') \exp\left\{-j\,\frac{k(x-x')^2}{2z}\right\} dx' \; . \tag{5.15}$$

It will be assumed that the amplitude of the aperture field is non-negligible over the region $\pm\, a/2$ of the aperture plane extending over many wavelengths, and that the phase variation is roughly symmetrical across the aperture, such that the corresponding angular spectrum is sufficiently narrow and centred on the direction of the positive z-axis for equation (5.15) to apply. Expanding the exponential in the integrand of equation (5.15),

$$g(x) = \sqrt{\frac{j\lambda}{z}} \exp\left\{-jk\left(z + \frac{x^2}{2z}\right)\right\} \cdot \frac{1}{\lambda} \int_{-a/2}^{a/2} f(x') \cdot$$

$$\exp\left(-j\frac{kx'^2}{2z}\right) \exp\left(jk\frac{x}{z}x'\right) dx' \quad . \tag{5.61}$$

The integrand here contains a parabolic phase term whose maximum value

$$\psi_{max} = \frac{k}{2z}\left(\frac{a}{2}\right)^2 \tag{5.62}$$

occurs at the extremities of the finite aperture. For a fixed aperture width and wavelength this tends to zero as z tends to infinity. We arbitrarily adopt the condition that the parabolic phase term is sensibly zero for

$$\psi_{max} \leqslant \pi/8 \tag{5.63}$$

and corresponds to

$$z \geqslant \frac{2a^2}{\lambda} \tag{5.64}$$

for the field over the plane z to have the same form as the field over a plane at infinity. Inequality (5.64) is known as **Rayleigh's far-field criterion.**

The validity of the above argument is most easily seen for $f(x') = 1$, when the integral of the equation (5.61) for a point on the axis ($x = 0$) reduces to a Fresnel integral. Inspection of Cornu's spiral, Fig. 5.5, and the subsequent exercise shows that the error in assessing the on-axis amplitude by taking ψ_{max} as zero rather than $\pi/8$ is only 0.69%. In practice aperture-field amplitudes are often tapered towards the edges of the aperture in order to reduce the sidelobe levels, which only serves to strengthen the argument.

For two-dimensional aperture-fields (see section 5.5.3) Rayleigh's far-field criterion is applied using the largest aperture dimension.

5.3.2 Field in the Focal Plane of a Focused System

A similar argument applies when the field of a radiating aperture is focused on to a plane a distance z_o away. (This can be achieved in practice by using a solid or artificial-dielectric lens, or by moving the feed of a parabolic-reflector antenna slightly away from the reflector).

Fig. 5.10 shows that the action of a focusing device is to introduce curvature of the wavefront centred on the focal point. This means that, for $z_o \gg a$, the

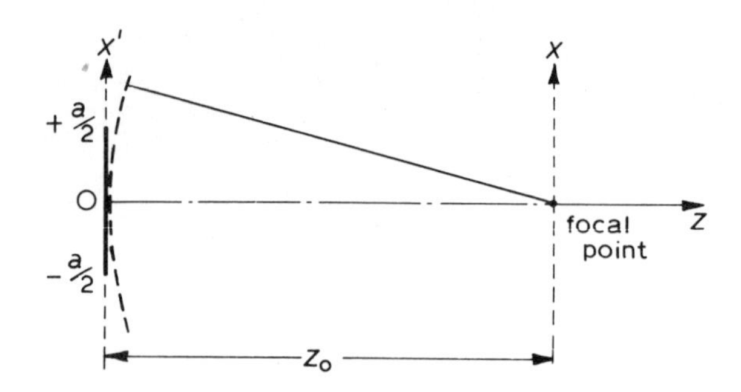

Fig. 5.10 – Focused aperture field.

aperture field $f(x')$ becomes $f(x')\exp\{+jkx'^2/(2z_o)\}$ and the tangential electric field over the focal plane, from equation (5.61), is

$$g(x) = \sqrt{\frac{j\lambda}{z_o}} \exp\left\{-jk\left(z_o + \frac{x^2}{2z_o}\right)\right\} F\left(\frac{x}{z_o}\right) \quad , \tag{5.65}$$

where

$$F\left(\frac{x}{z_o}\right) = \frac{1}{\lambda} \int_{-a/2}^{a/2} f(x') \exp\left(jk\frac{x}{z_o}x'\right) \mathrm{d}x' \tag{5.66}$$

is the Fourier transform of the original, unfocused, aperture field.

This well-known result in optics has been used as a basis for finding the far field of an antenna when it is otherwise difficult to satisfy the far-field criterion, owing to largeness of antenna dimensions or a restricted measurement site (see Johnson, Ecker and Hollis (1973)).

5.4 DETAILED STRUCTURE OF THE RADIATED FIELD

It is of the very essence of diffraction that electromagnetic waves spread out as they propagate through a uniform medium. We have already seen how an aperture field, whose phase is constant over the aperture plane, eventually becomes a spherical wave in the far field. We have also seen that the Rayleigh distance provides a convenient practical measure of the onset of the far field. In this section we will examine the radiated field in somewhat greater detail, using Fresnel's diffraction formula, for the two cases of a uniform and a Gaussian-shaped aperture field. This will reveal that there is a region close to the aperture, at distances much less than the Rayleigh distance, in which the radiated field is approximately collimated. The analysis also provides a check on the Rayleigh criterion, which was established by reference only to the field on axis, so the question remains as to how closely the field as a whole approximates the far field.

5.4.1 Radiation from a uniform aperture field

The simplest approach to this (two-dimensional) problem is to imagine a plane wave normally incident on a slit of width a cut in a thin, perfectly conducting sheet as shown in Fig. 5.11. Provided this slit is a reasonable number of wave-

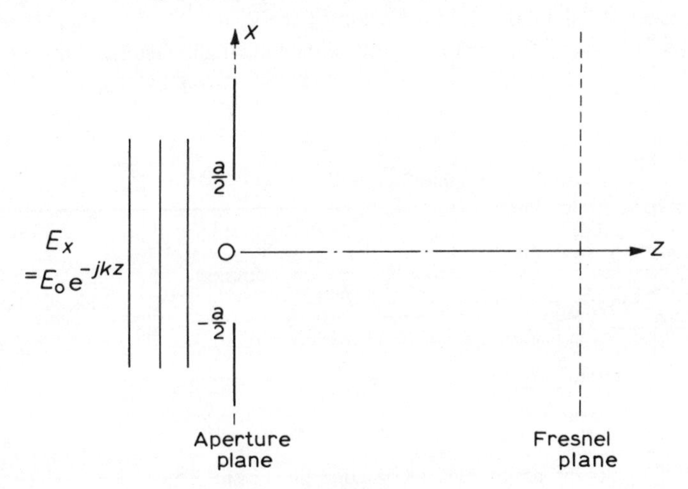

Fig. 5.11 – A slit of width a in a perfectly conducting sheet normally illuminated by a plane wave.

lengths wide, the aperture field can be taken to be the constant E_o over the slit and zero outside, that is

$$E_{ax}(x) = E_o \operatorname{rect}_a(x) \quad . \tag{5.67}$$

The tangential electric field over the Fresnel plane, at any z more than a few wavelengths distant from the aperture plane, is given by substituting this aperture field into Fresnel's diffraction formula (5.15), and is

$$E_x(x,z) = E_o \sqrt{\frac{j}{\lambda z}} \exp(-jkz) \int_{-a/2}^{a/2} \exp\left\{-jk\frac{(x-x')^2}{2z}\right\} dx' \quad . \tag{5.68}$$

The solution to this can be written down immediately in terms of Fresnel integrals (equation 5.31) as

$$E_x(x,z) = E_o \sqrt{\frac{j}{2}} \exp(-jkz) \left\{\mathscr{F}\left[\sqrt{\frac{2}{\lambda z}}\,(-x + a/2)\right] - \mathscr{F}\left[\sqrt{\frac{2}{\lambda z}}\,(-x - a/2)\right]\right\} \quad . \tag{5.69}$$

But initially it is more instructive to build up a physical picture of the radiation combining the diffraction from two knife edges, using the results of section 5.2. In Fig. 5.12 two lines inclined at angles $\pm \sin^{-1}(\lambda/a)$ to the z-axis give the directions to the first nulls of the far field pattern (cf. equation (2.58)). Since λ/a is assumed small, the projection of the aperture of width a on to these lines occurs at $z = a^2/(2\lambda)$, which is just a quarter of the distance we have taken

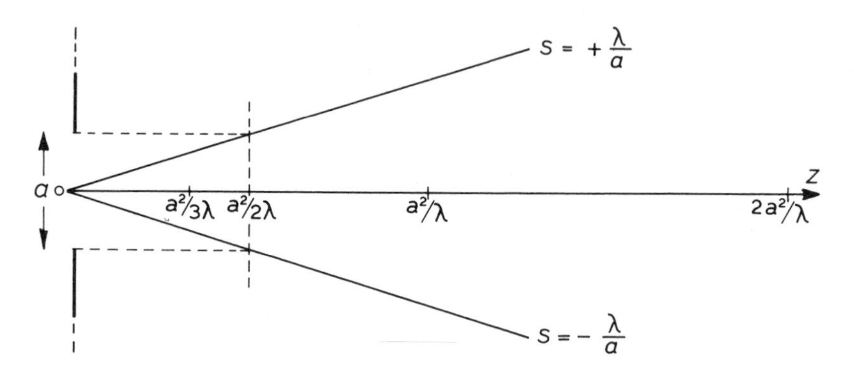

Fig. 5.12 – Salient points used to describe the Fresnel region from a uniform aperture field distribution.

to be the Rayleigh distance. The point $z = a^2/(3\lambda)$ is approximately where the first maxima of the two knife-edge diffraction patterns coincide, when the Fresnel variable $\nu = 1.25$ (Fig. 5.5). The field dies away reasonably quickly on each side of the projected aperture. At distances nearer than this to the aperture the Fresnel parameter $\sqrt{(\lambda z)}$ becomes smaller and smaller, the knife-edge pattern fringes crowd closer together, and the Fresnel-plane pattern approximates that of the rectangular aperture distribution. While at the Rayleigh distance and beyond, the Fresnel-plane pattern is expected to have the sinc-function form of the far field.

Fig. 5.13 shows computations, based on equation (5.69), of the magnitude of the tangential electric field across Fresnel planes (a) near to the aperture ($z = a^2/(30\lambda)$), (b) towards the end of the collimated region ($z = a^2/(3\lambda)$), and (c) at the Rayleigh distance ($z = 2a^2/\lambda$). They bear out the form of the fields anticipated above, with (a) being of near rectangular form except in the vicinity of $x = \pm a/2$, (b) rather shapeless but largely confined to the projected aperture, and (c) of the form of the far field, with the first zeros occurring at $x = \pm 2a$, as they should at the Rayleigh distance.

Exercise

Use the asymptotic form (equation (5.38)) for the Fresnel integral to show analytically that the off-axis form of the Fresnel field at the Rayleigh distance is approximately a sinc function. Check that the first zeros occur at $x = \pm 2a$,

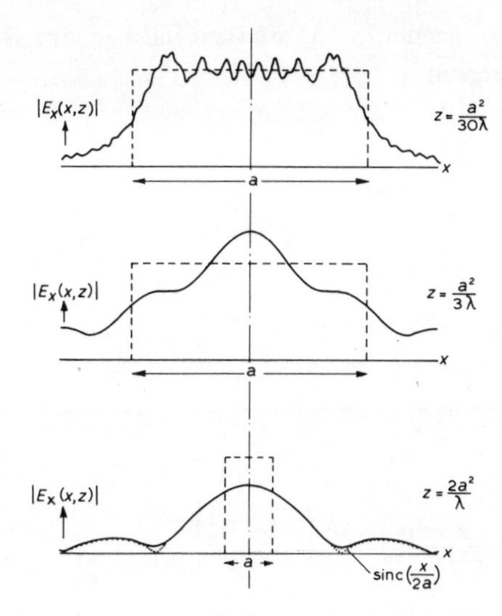

Fig. 5.13 – Fresnel region field from a uniformly illuminated aperture of width a.

and that the first sidelobe level differs by a few percent in amplitude from the far-field value.

5.4.2 Radiation from a Gaussian-shaped aperture field

A useful model for an antenna is one with a Gaussian-shaped aperture field, whose far-field pattern has no sidelobes (see Exercise, p. 63), since it is again Gaussian. The fundamental mode of the field emitted by a laser has this form. We will now show that the Gaussian shape of the magnitude of the lateral field distribution is maintained right from the aperture, through the Fresnel region into the far field.

Suppose that the one-dimensional aperture field is

$$E_{ax}(x) = E_o \exp\left(-\frac{x^2}{{w_o}^2}\right) , \tag{5.70}$$

which is of width $2w_o$ between the $1/e$ points and is of uniform phase. Its angular spectrum is

$$F(s) = \frac{\sqrt{\pi w_o}}{\lambda} E_o \exp\left\{-\frac{\pi^2 w_o^2 s^2}{\lambda^2}\right\} \tag{5.71}$$

whose angular width between $1/e$ points will be small if w_o is many wavelengths, and is then approximately $2\lambda/(\pi w_o)$. The conditions for the application of

Fresnel's diffraction formula (5.15) are then fulfilled, and the tangential electric field in the Fresnel region is

$$E_x(x,z) = \sqrt{\frac{j}{\lambda z}}\, E_o \exp(-jkz)$$

$$\int_{-\infty}^{\infty} \exp\left\{-\frac{x'^2}{w_o^2} - jk\,\frac{(x-x')^2}{2z}\right\} dx' \;, \tag{5.72}$$

which has the solution, using the standard integral (2.96),

$$E_x(x,z) = E_o \sqrt{\frac{j\pi}{\lambda z}} \exp\left\{-jk\left(z + \frac{x^2}{2z}\right)\right\}$$

$$\cdot \frac{1}{\sqrt{\dfrac{1}{w_o^2} + \dfrac{jk}{2z}}} \exp\left\{-\frac{\dfrac{k^2x^2}{4z^2}}{\dfrac{1}{w_o^2} + \dfrac{jk}{2z}}\right\} . \tag{5.73}$$

This can be rearranged to give

$$E_x(x,z) = E_o \sqrt{\frac{j\,\dfrac{\pi w_o^2}{\lambda z}}{1 + j\,\dfrac{\pi w_o^2}{\lambda z}}} \exp\left\{-jk\left[z + \frac{x^2}{2R(z)}\right]\right\} \exp\left\{-\frac{x^2}{w^2(z)}\right\} , \tag{5.74}$$

where the 'half waist'

$$w(z) = w_o \sqrt{1 + \left(\frac{\lambda z}{\pi w_o^2}\right)^2} \tag{5.75}$$

and the radius of curvature of the wave front

$$R(z) = z\left[1 + \left(\frac{\pi w_o^2}{\lambda z}\right)^2\right] \tag{5.76}$$

both depend on z.

The character of the Fresnel region field is now clear from equation (5.74). The lateral (x) dependence of the amplitude of the field is Gaussian at any distance z from the aperture (see Fig. 5.14). The $1/e$ points occur at $\pm w(z)$, which starts from its value w_o at the aperture and eventually becomes asymptotic to the radial lines from the centre of the aperture inclined at $s = \pm\lambda/(\pi w_o)$ to the z-axis in the far field. The radius of the wavefront is infinite at the aperture,

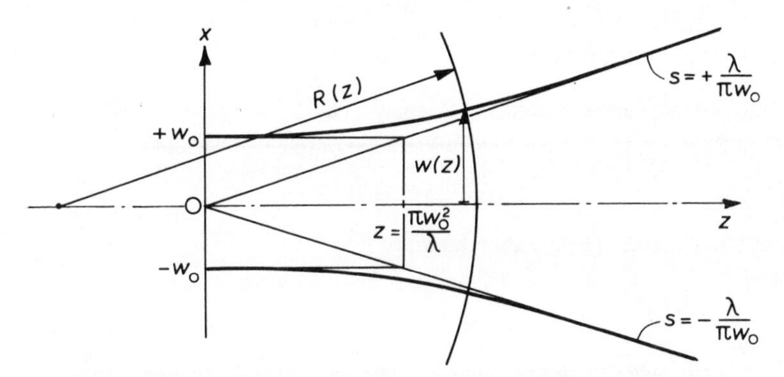

Fig. 5.14 – Trajectories of the $1/e$ points ($\pm\ w(z)$) of the amplitude of the Fresnel region field, and the associated wavefront curvature of radius $R(z)$.

then gradually decreases to become equal to z in the far field. The radiation from a Gaussian-shaped aperture field is clearly in the form of a beam, initially approximately collimated. A transition occurs in the neighbourhood of

$$\frac{\pi w_o^2}{\lambda z} = 1 \quad , \quad \text{or } z = \frac{\pi w_o^2}{\lambda} \tag{5.77}$$

which, though memorable, cannot be taken to be the Rayleigh distance appropriate to this case. Beyond this point the beam begins to spread more noticeably, eventually acquiring the far-field character of a cylindrically spreading wave centered on the aperture.

Exercise

In order to apply the Rayleigh criterion, take the 'effective width' a of the Gaussian aperture field to be (a) $4w_o$ and (b) $6w_o$ and compare the field structure with that of the far field in order to decide which gives an adequate far-field criterion.

5.5 FRESNEL DIFFRACTION IN THREE DIMENSIONS

5.5.1 Fresnel's diffraction formula

It was shown in section 3.5.1 that the electromagnetic field radiating into the half-space $z \geqslant 0$ is completely and precisely represented in terms of two angular plane-wave spectra $F_x(\alpha,\beta)$ and $F_y(\alpha,\beta)$. Typically

$$E_x(x,y,z) = \int_{-\infty}^{\infty}\int_{-\infty}^{\infty} F_x(\alpha,\beta) \exp\{-jk(\alpha x + \beta y + \gamma z)\}\, d\alpha d\beta \tag{5.78}$$

for an x-polarized tangential electric aperture field, over the plane $z = 0$, of $E_{ax}(x,y) = E_x(x,y,0)$, where

$$F_x(\alpha,\beta) = \frac{1}{\lambda^2}\int_{-\infty}^{\infty}\int_{-\infty}^{\infty} E_{ax}(x,y) \exp\{+jk(\alpha x + \beta y)\}\, dxdy \quad . \tag{5.79}$$

The direction cosines (α,β,γ) are related by

$$\alpha^2 + \beta^2 + \gamma^2 = 1 \tag{5.80}$$

and $\gamma = \cos\theta$, where θ is the angle a particular direction makes with the z-axis.

The fundamental approximation of Fresnel diffraction theory (cf. section 5.1) is that the angular spectrum is negligible outside a narrow range of angles centred on the direction of the z-axis. Hence, using the approximation that

$$\gamma = \sqrt{(1 - \alpha^2 - \beta^2)} \approx 1 - \tfrac{1}{2}(\alpha^2 + \beta^2) \quad , \tag{5.81}$$

equation (5.78) can be written approximately as

$$E_x(x,y,z) = \exp(-jkz)\int_{-\infty}^{\infty}\int_{-\infty}^{\infty} F_x(\alpha,\beta)$$

$$\exp\left\{+j\frac{k}{2}(\alpha^2+\beta^2)z\right\} \exp\{-jk(\alpha x + \beta y)\}\, d\alpha d\beta \tag{5.82}$$

for which we require the convolution theorem for two-dimensional Fourier transforms.

Exercise

Given the following two-dimensional Fourier transform pairs

$$E(x,y) \longleftrightarrow F(\alpha,\beta)$$

and

$$H(x,y) \longleftrightarrow G(\alpha,\beta)$$

show that

$$\frac{1}{\lambda^2}\int_{-\infty}^{\infty}\int_{-\infty}^{\infty} E(x',y')\,H(x-x',\,y-y')\,\mathrm{d}x'\mathrm{d}y'$$

$$\longleftrightarrow F(\alpha,\beta)\,G(\alpha,\beta) \quad . \tag{5.83}$$

It then follows (cf. equation (5.14)) that

$$E_x(x,y,z) = \frac{j}{\lambda z}\exp(-jkz)\int_{-\infty}^{\infty}\int_{-\infty}^{\infty} E_{ax}(x',y')$$

$$\exp\left\{-\frac{jk}{2z}[(x-x')^2+(y-y')^2]\right\}\,\mathrm{d}x'\mathrm{d}y' \quad , \tag{5.84}$$

which is Fresnel's diffraction formula for three-dimensional fields, valid under the restriction that the angular spectrum $F_x(\alpha,\beta)$ is narrowly confined about the direction $\theta = 0$.

Exercise
Show that equation (5.84) satisfies the three-dimensional form of the parabolic equation approximation (cf. equation (5.24))

$$\frac{\partial^2 u}{\partial x^2}+\frac{\partial^2 u}{\partial y^2}-j2k\frac{\partial u}{\partial z}=0 \tag{5.85}$$

with

$$u(x,y,z) = E_x(x,y,z)\exp(-jkz) \quad . \tag{5.86}$$

Having formally derived Fresnel's diffraction formula under the restriction that it applies to fields with narrow angular spectra, it must be pointed out that this restriction is not always strictly required in practice, as the following example shows.

5.5.2 Diffraction by a conducting half plane

Suppose that a conducting half plane intrudes a height h above the line joining a point transmitter to a point receiver, as in Fig. 5.15. The transmitter can be imagined to have aperture dimensions $a \times b$ very small compared to the wavelength. Then

$$E_{ax}{}^{\mathrm{T}}(x,y) = E_0\delta(x)\,\delta(y)\,ab \tag{5.87}$$

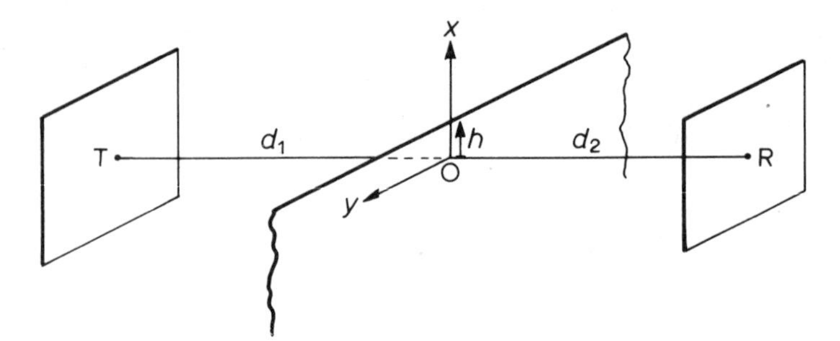

Fig. 5.15 – Diffracting knife edge, a distance d_1 from the transmitter T and d_2 from the receiver R.

and the transmitter angular spectrum

$$F_x{}^{\mathrm{T}}(\alpha,\beta) = \frac{E_0 ab}{\lambda^2} \tag{5.88}$$

is independent of direction. The transmitted field is then (cf. equation (3.61))

$$E_x{}^{\mathrm{T}}(r,\theta,\phi) = \frac{j\lambda}{r} \exp(-jkr) \cos\theta \, F_x(\alpha,\beta) \tag{5.89}$$

which will be approximated at the plane $z = d_1$ of the knife edge as

$$E_x{}^{\mathrm{T}}(x,y,d_1) = \frac{jE_0 ab}{\lambda d_1} \exp(-jkd_1) \exp\left\{-jk \frac{x^2+y^2}{2d_1}\right\} \quad . \tag{5.90}$$

This is a good approximation only in the vicinity of the axis. But, as will be apparent, this is the only part of the incident field that contributes significantly to the final result.

The aperture field just to the right of the knife edge will be taken to be

$$E_{ax}{}^{K}(x,y) = E_x{}^{\mathrm{T}}(x,y,d_1)\, U(x-h) \tag{5.91}$$

where $U(\,)$ is Heaviside's unit step function. Applying Fresnel's diffraction formula (5.84), the field over the plane of the receiver is

$$E_x{}^{\mathrm{R}}(x,y,d_2) = -\frac{E_0 ab}{\lambda^2 d_1 d_2} \exp\left\{-jk(d_1+d_2)\right\}$$

$$\int_h^\infty dx' \int_{-\infty}^\infty dy' \exp\left\{-\frac{jk}{2}\left[\frac{x'^2+y'^2}{d_1} + \frac{(x-x')^2+(y-y')^2}{d_2}\right]\right\}. \tag{5.92}$$

At the receiver itself ($x = 0, y = 0$) the received field is

$$E_x{}^{R}(0,0,d_2) = -\frac{E_0 ab}{\lambda^2 d_1 d_2} \exp\left\{-jk(d_1 + d_2)\right\}$$

$$\int_h^{\infty} dx' \int_{-\infty}^{\infty} dy' \exp\left\{-\frac{jk}{2}\left(\frac{1}{d_1} + \frac{1}{d_2}\right)(x'^2 + y'^2)\right\} \quad (5.93)$$

The y'-integration can be carried out using the standard integral of equation (2.96), and the x'-integration using Fresnel integrals (equation (5.31)). Thus

$$E_x{}^{R}(0,0,d_2) = \frac{E_0 ab}{\lambda}\sqrt{\frac{j}{2}}\, \frac{\exp\{-jk(d_1 + d_2)\}}{d_1 + d_2}$$

$$\left[\mathscr{F}(\infty) - \mathscr{F}\left(h\sqrt{\frac{2(d_1 + d_2)}{\lambda d_1 d_2}}\right)\right] \quad (5.94)$$

To check this formula, remove the screen to $h = -\infty$, which gives the unobstructed field as

$$E_{xo}{}^{R}(0,0,d_2) = \frac{jE_0 ab}{\lambda(d_1 + d_2)} \exp\left\{-jk(d_1 + d_2)\right\} \quad , \quad (5.95)$$

in agreement with equation (5.90). The ratio of obstructed to unobstructed received field is therefore

$$\frac{E_x{}^{R}(0,0,d_2)}{E_{xo}{}^{R}(0,0,d_2)} = \frac{1}{\sqrt{2j}}\left[\mathscr{F}(\infty) - \mathscr{F}\left(h\sqrt{\frac{2(d_1 + d_2)}{\lambda d_1 d_2}}\right)\right] \quad . \quad (5.96)$$

Note that the Fresnel parameter here is $\sqrt{(\lambda D)}$, where

$$\frac{1}{D} = \frac{1}{d_1} + \frac{1}{d_2} \quad . \quad (5.97)$$

In order to justify the seemingly gross approximation used for the field incident on the knife edge (equation (5.90)) it should be observed that the integrand of equation (5.93) is a pure phase term that depends on the squares of both integration variables x' and y'. As x' and y' increase the integrand revolves more and more rapidly round the unit circle, with the result that adjacent contributions to the integral cancel, leaving the axial contribution as dominant. (This is the phenomenon of Fresnel zones, to be discussed later in section (5.5.4)). The same argument applies even when the amplitude of the integrand varies with x' and y', provided the amplitude variations are slower than the oscillations in the phase. This implies that the knife edge must be in the far field of the transmitter for the above analysis to apply.

A more general and precise analysis is available based on the antenna coupling formula of section 4.3. It applies to arbitrary transmitting and receiving patterns, and there is no restriction on the position of the knife edge. It is correspondingly more complicated, so only the first few steps will be given.

Referring again to Fig. 5.15, if the transmitted angular spectrum is $F_x{}^{\mathrm{T}}(\alpha,\beta)$ the field incident on the plane containing the knife edge is (without approximation)

$$E_x{}^{\mathrm{T}}(x,y,d_1) = \int_{-\infty}^{\infty}\int_{-\infty}^{\infty} F_x{}^{\mathrm{T}}(\alpha,\beta) \exp\{-jk(\alpha x + \beta y + \gamma d_1)\}\, d\alpha d\beta \quad . \tag{5.98}$$

The aperture field just beyond the knife edge is, to a good approximation,

$$E_{ax}{}^{\mathrm{K}}(x,y) = E_x{}^{\mathrm{T}}(x,y,d_1)\, U(x-h) \quad . \tag{5.99}$$

The angular spectrum of the field radiating into the region to the right of the knife edge is

$$F_x{}^{\mathrm{K}}(\alpha,\beta) = \frac{1}{\lambda^2}\int_{-\infty}^{\infty}\int_{-\infty}^{\infty} E_{ax}{}^{\mathrm{K}}(x,y) \exp\{jk(\alpha x + \beta y)\}\, dx dy \quad . \tag{5.100}$$

Substituting (5.98) and (5.99) into equation (5.100) gives

$$F_x{}^{\mathrm{K}}(\alpha,\beta) = \frac{1}{\lambda^2}\int_{h}^{\infty} dx \int_{-\infty}^{\infty} dy \int_{-\infty}^{\infty} d\alpha' \int_{-\infty}^{\infty} d\beta'\, F_x{}^{\mathrm{T}}(\alpha',\beta') \exp(-jk\gamma' d_1)$$
$$\exp\{jk[(\alpha-\alpha')x + (\beta-\beta')y]\} \tag{5.101}$$

which is an angular spectrum referred to reference point O on the knife edge.

The coupling formula of equation (4.57) for a receiver, whose transmitting pattern is defined by the angular spectrum $F_x{}^{\mathrm{R}}(\alpha,\beta)$, at the coordinate point $(0,0,d_2)$ is

$$c = \frac{\lambda^2}{2Z}\int_{-\infty}^{\infty}\int_{-\infty}^{\infty} F_x{}^{\mathrm{K}}(\alpha,\beta)\, F_x{}^{\mathrm{R}}(-\alpha,\beta)\, \frac{1-\beta^2}{\gamma} \exp(-jk\gamma d_2)\, d\alpha d\beta$$

in which Z is the characteristic impedance of the medium. Substituting equation (5.101), the coupled signal is

$$c = \frac{1}{2Z}\int_{-\infty}^{\infty} d\alpha \int_{-\infty}^{\infty} d\beta \int_{h}^{\infty} dx \int_{-\infty}^{\infty} dy \int_{-\infty}^{\infty} d\alpha' \int_{-\infty}^{\infty} d\beta' \, F_x{}^{\mathrm{T}}(\alpha',\beta')$$

$$F_x{}^{\mathrm{R}}(-\alpha,\beta)\,\frac{1-\beta^2}{\gamma}\,\exp\{-jk\,(\gamma' d_1 + \gamma d_2)\}$$

$$\exp\{+jk\,[(\alpha-\alpha')\,x + (\beta-\beta')\,y]\} \quad . \tag{5.102}$$

The x-integration is (cf. equation (5.26))

$$\int_{h}^{\infty} \exp\{jk\,(\alpha-\alpha')\,x\}\,dx$$

$$= \frac{\lambda}{2}\exp\{jk\,(\alpha-\alpha')\,h\}\left[\delta(\alpha-\alpha') + \frac{j}{\pi(\alpha-\alpha')}\right] , \tag{5.103}$$

and the y-integration yields a simple delta function.

Writing the coupled signal as

$$c = c_1 + c_2 \tag{5.104}$$

we have that

$$c_1 = \frac{\lambda^2}{4Z}\int_{-\infty}^{\infty} d\alpha \int_{-\infty}^{\infty} d\beta \int_{-\infty}^{\infty} d\alpha' \int_{-\infty}^{\infty} d\beta' \, F_x{}^{\mathrm{T}}(\alpha',\beta')\, F_x{}^{\mathrm{R}}(-\alpha,\beta)$$

$$\frac{1-\beta^2}{\gamma}\exp\{-jk\,(\gamma' d_1 + \gamma\, d_2)\}$$

$$\exp\{jk\,(\alpha-\alpha')\,h\}\,\delta(\alpha-\alpha')\,\delta(\beta-\beta') \tag{5.105}$$

which is precisely half the signal that would have been coupled from T to R had there been no obstructing knife edge.

The remaining part of the coupled signal

$$c_2 = \frac{\lambda^2}{4Z}\int_{-\infty}^{\infty} d\alpha \int_{-\infty}^{\infty} d\beta \int_{-\infty}^{\infty} d\alpha' \int_{-\infty}^{\infty} d\beta' \, F_x{}^{\mathrm{T}}(\alpha',\beta')\, F_x{}^{\mathrm{R}}(-\alpha,\beta)$$

$$\frac{1-\beta^2}{\gamma} \exp\{-jk(\gamma' d_1 + \gamma\, d_2)\}$$

$$\exp\{jk\,(\alpha-\alpha')\,h\}\ \frac{j}{\pi(\alpha-\alpha')}\ \delta(\beta-\beta') \tag{5.106}$$

needs to be evaluated in terms of Cauchy principal values. (A useful reference to the latter is Dennery and Krzwicki (1967) p. 61).

5.5.3 Transition to the far field

Applying Fresnel's diffraction formula to a rectangular aperture

$$E_x(x,y,z) = \frac{j}{\lambda z} \exp(-jkz) \int_{-a/2}^{a/2} \mathrm{d}x' \int_{-b/2}^{b/2} \mathrm{d}y'\, E_{ax}(x',y')$$

$$\exp\left\{-\frac{jk}{2z}[(x-x')^2 + (y-y')^2]\right\} \quad , \tag{5.107}$$

it is clear, using the same arguments that were employed in section 5.3, that the onset of the far field is determined by whichever is the larger of a or b. The Rayleigh far field criterion is therefore

$$z \geqslant \frac{2[\max(a,b)]^2}{\lambda} \tag{5.108}$$

A uniformly illuminated circular aperture provides the opportunity of comparing the on-axis gain at various distances, such as the Rayleigh distance, with its far-field value. Supposing that the aperture field $E_{ax}(x,y) = E_0$ is constant over a circular aperture of radius a and zero outside, Fresnel's diffraction formula can be rewritten in terms of polar (ρ,ϕ) aperture coordinates to give the field at a point z on the axis as

$$E_x(0,0,z) = \frac{jk}{z} \exp(-jkz) \int_0^a \exp\left\{-\frac{jk\rho^2}{2z}\right\} \rho \mathrm{d}\rho \tag{5.109}$$

Integration yields

$$E_x(0,0,z) = 2jE_0 \exp\left\{-jk\left(z + \frac{a^2}{4z}\right)\right\} \sin\left(\frac{ka^2}{4z}\right) \tag{5.110}$$

The only other field component is that of the magnetic field which is approximately

$$H_y(0,0,z) = \frac{1}{Z}\, E_x(0,0,z) \tag{5.111}$$

as will be confirmed later (see equation (5.130)). The power flow is therefore in the axial direction, that is,

$$S_z(0,0,z) = \frac{|E_x(0,0,z)|^2}{2Z} = \frac{2|E_o|^2}{Z}\,\sin^2\left(\frac{ka^2}{4z}\right) \tag{5.112}$$

The total power flowing through the aperture is

$$P_o = \pi a^2 \frac{|E_o|^2}{2Z} \tag{5.113}$$

and the (on-axis) gain is therefore

$$G = \frac{4\pi z^2 S_z}{P_o} \tag{5.114}$$

which is, from equations (5.112) and (5.113)

$$G = \frac{4\pi(\pi a^2)}{\lambda^2} \cdot \frac{\sin^2\left(\dfrac{ka^2}{4z}\right)}{\left(\dfrac{ka^2}{4z}\right)^2} \quad . \tag{5.115}$$

And in terms of the far-field gain G_o of a uniformly illuminated aperture (equation (4.20)) of area πa^2,

$$\frac{G}{G_o} = \left[\frac{\sin\left(\dfrac{ka^2}{4z}\right)}{\dfrac{ka^2}{4z}}\right]^2 \tag{5.116}$$

which has the value of 0.987 at the Rayleigh distance $z = 8a^2/\lambda$, for example.

5.5.4 Fresnel Zones

The following construction, due to Fresnel, is useful in showing how the field arising from an extensive aperture can be thought of as being due just to a small part, known as the **first Fresnel zone.** We will treat a plane aperture, the x-y plane, across which the field will be taken to be uniform. The Fresnel zones are separated by circles (see Fig. 5.16) which are formed by the intersection with

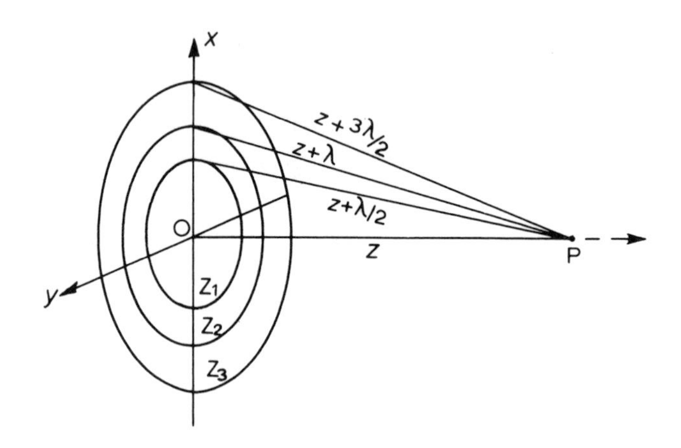

Fig. 5.16 – Construction of Fresnel zones on the x-y plane for the observation point P.

the aperture plane of spheres, centred on the point of observation P(0,0,z) whose radii are succesively $z + \lambda/2$, $z + 2(\lambda/2)$, $z + 3(\lambda/2)$, and so on. The first Fresnel zone is therefore a circle while the second and higher zones are annuli.

If z is much larger than λ the radius of the first Fresnel zone is $(\lambda z)^{1/2}$. Inserting this into equation (5.110) gives

$$E_{x1}(0,0,z) = 2E_o \exp(-jkz) \tag{5.117}$$

which is precisely twice the value of $E_x(0,0,z)$ had the aperture been infinite, and the incident plane wave unobstructed. We will now examine this seemingly curious result in more detail, using a vector form of Huygens' principle (see section 3.5.2).

Consider an element $\mathrm{d}x\mathrm{d}y$ of a wavefront, situated at the origin in the x-y plane, whose electric field is x-polarized and of complex amplitude E_o. Its angular spectrum is

$$F_x(\alpha,\beta) = \frac{E_o}{\lambda^2}\ \mathrm{d}x\mathrm{d}y \tag{5.118}$$

and the radiated electric field (equation (3.61)) is therefore

$$\mathbf{dE}(r,\theta,\phi) = j2\pi \frac{E_o}{\lambda^2} \frac{\exp(-jkr)}{kr} [\mathbf{u}_x \cos\theta - \mathbf{u}_z \sin\theta \cos\phi] \, dx dy \tag{5.119}$$

Applying this result to the present situation by shifting the radiating element to an arbitrary point in the x-y plane, as in Fig. 5.17, the elemental field at $P(0,0,z)$ is

$$\mathbf{dE}(0,0,z) = j2\pi \frac{E_o}{\lambda^2} \frac{\exp(-jkr)}{kr} [\mathbf{u}_x \cos\theta + \mathbf{u}_z \sin\theta \cos\phi] \, dx dy \tag{5.120}$$

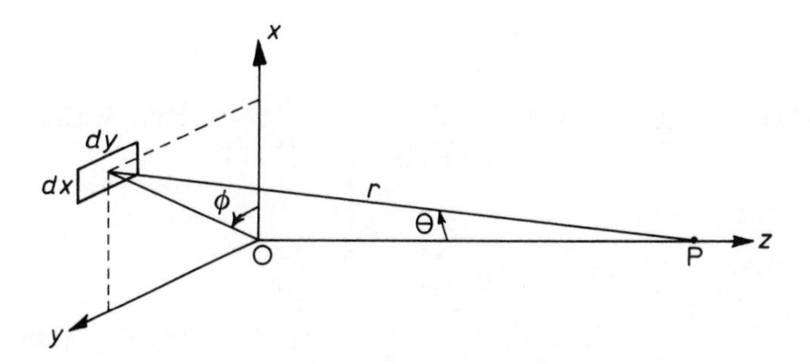

Fig. 5.17 – Radiating element in the x-y plane.

(note that θ is unchanged but that ϕ corresponds to $\phi + \pi$ in equation (5.119)). Changing to polar coordinates and integrating round an annulus of radius ρ and width $d\rho$,

$$\mathbf{dE}(0,0,z) = \int_0^{2\pi} d\phi \, j\, 2\pi \frac{E_o}{\lambda^2} \frac{\exp(-jkr)}{kr} [\mathbf{u}_x \cos\theta + \mathbf{u}_z \sin\theta \cos\phi] \, \rho d\rho$$

$$= \mathbf{u}_x \, jk^2 E_o \cos\theta \frac{\exp(-jkr)}{kr} \rho d\rho \quad , \tag{5.121}$$

which is purely x-polarized.

Now integrating over the ith Fresnel zone which involves changing to r as the integration variable over the range from $z + (i\text{-}1)(\lambda/2)$ to $z + i(\lambda/2)$, the electric field at P due to the ith zone is

$$E_{xi}(0,0,z) = jkE_o \int_{z + (i-1)\lambda/2}^{z + i\lambda/2} \frac{z}{r} \exp(-jkr)\, dr \quad . \tag{5.122}$$

Make the reasonable approximation that $z/r = C_i$ is constant over each zone, so that

$$\begin{aligned} E_{xi}(0,0,z) &= jkE_o\, C_i \frac{1}{-jk} \left[\exp\left\{-jk\left[z + i(\lambda/2)\right]\right\} \right. \\ &\quad \left. - \exp\left\{-jk\,[z + (i-1)(\lambda/2)]\right\}\right] \\ &= 2E_o \exp(-jkz)\,(-1)^{i+1} C_i \end{aligned} \tag{5.123}$$

in agreement with equation (5.117) for the first Fresnel zone, with $C_1 = 1$.

Then summing, to give the total electric field at P,

$$E_x(0,0,z) = 2E_o \exp(-jkz) \sum_{i=1}^{\infty} (-1)^{i+1} C_i \tag{5.124}$$

The summation (see Born and Wolf (1970) p. 372)

$$\sum = C_1 - C_2 + C_3 - C_4 + C_5 - \ldots \tag{5.125}$$

can be written

$$\begin{aligned} \sum &= \frac{C_1}{2} + \left(\frac{C_1}{2} - C_2 + \frac{C_3}{2}\right) + \left(\frac{C_3}{2} - C_4 + \frac{C_5}{2}\right) + \ldots \\ &\approx \frac{C_1}{2} \approx \frac{1}{2} \end{aligned} \tag{5.126}$$

since the bracketed terms in the summation are all approximately zero, and $C_n \to 0$ as $n \to \infty$. Thus the field on axis

$$E_x(0,0,z) = E_o \exp(-jkz) = \tfrac{1}{2} E_{x1}(0,0,z) \tag{5.127}$$

arising from the entire uniformly illuminated aperture plane is just one half of the field at that point due to the first Fresnel zone.

Similar arguments give the magnetic field at P due to the wavefront element in Fig. 5.17 (cf. equations (3.94) and (3.62)) as

$$\begin{aligned} d\mathbf{H}(0,0,z) = j2\pi \frac{E_o}{Z\lambda^2} \frac{\exp(-jkr)}{kr} [& -\mathbf{u}_x \sin^2\theta \sin\phi \cos\phi \\ & + \mathbf{u}_y (1 - \sin^2\theta \sin^2\phi) \\ & + \mathbf{u}_z \sin\theta \cos\theta \cos\phi] \, dx dy \quad . \end{aligned} \tag{5.128}$$

When integrated around an annulus $\rho d\rho$ this becomes

$$d\mathbf{H}(0,0,z) = \mathbf{u}_y \, jk^2 \frac{E_o}{Z} (1 - \tfrac{1}{2}\sin^2\theta) \frac{\exp(-jkr)}{kr} \rho d\rho \tag{5.129}$$

which contains only a y component. Then, proceeding as for the electric field, the final result is

$$H_y(0,0,z) = \frac{E_o}{Z} \exp(-jkz) = \tfrac{1}{2} H_{y1}(0,0,z) \quad . \tag{5.130}$$

CHAPTER 6

Reflection from perfectly conducting surfaces

A vital part is played by reflecting metal surfaces in the performance of many aperture antennas. In the majority of the types cited in section 1.3 metal surfaces with a variety of contours are used to reshape and redirect the wavefront emanating from the primary waveguide source. Exact solutions are usually difficult to obtain, and even when they are available tend to be cumbersome and difficult to use. Approximate methods of analysis will therefore be developed which preserve the physical picture and also provide perfectly adequate engineering answers in most practical cases. The validity of this last point is particularly true in view of the uncertainties that arise at other stages of the analysis, such as the precise form of the source field, and the effects of the eventual obstruction of the emerging wave-field, by the primary source and its feeder waveguide or by a subreflector and its supports.

An infinite flat plane illuminated by a plane wave provides a basic model for our understanding of the problem. The analysis in this case is exact and will be treated first. Many of the details in the analysis will be of use in later sections of the chapter when it is established that finite surfaces, whose dimensions and radii of curvature are large compared to the wavelength, have a local effect on the relation between the incident and reflected fields which is approximately the same as that due to an infinite tangent plane at that point on the surface. For ease of application the analysis will be couched initially in vector form, free of any particular coordinate system.

6.1 REFLECTION BY AN INFINITE FLAT PLANE

If an ideal plane wave is incident on a perfectly conducting, infinite flat plane, the reflected field is a similar plane wave. The reason for this is that, in order to satisfy the boundary conditions at the conducting surface, the tangential component of the reflected electric field at the surface is essentially an aperture field of uniform amplitude of infinite lateral extent. The corresponding radiation away from the surface is therefore a plane wave (cf. equation (2.57)) travelling in a direction, determined by the phase at the surface of the incident wave, such that the angle of reflection is equal to the angle of incidence. This is known as the law of reflection. Fig. 6.1 shows the two principal cases when the incident plane wave is linearly polarized, (a) with the electric field directed parallel to the plane of incidence and (b) with the electric field perpendicular to the plane of incidence. The designations $\mathbf{e}^+$ and $\mathbf{e}^-$ for the two respective polarizations conveniently identify them as having the corresponding reflection coefficients of +1 and −1.

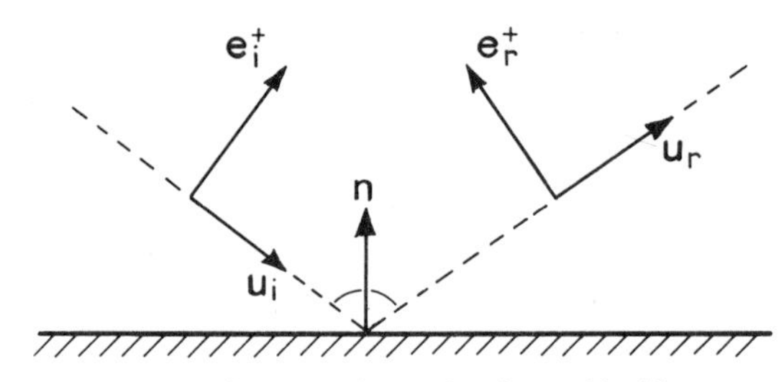

(a) Electric field parallel to the plane of incidence.

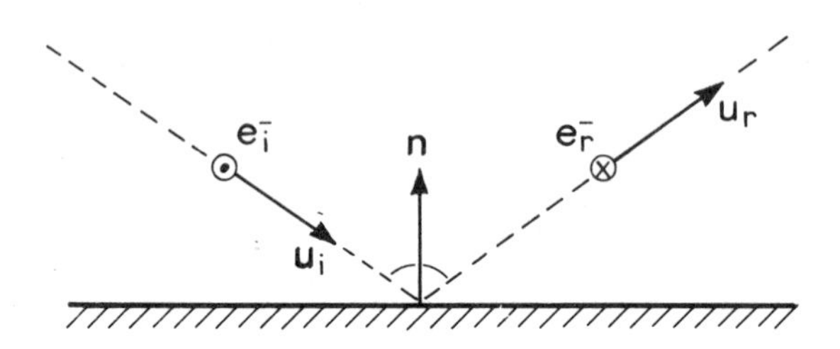

(b) Electric field perpendicular to the plane of incidence.

Fig. 6.1 – Electric field of a plane wave on reflection from a perfectly conducting infinite plane.

If the direction of the incident plane wave is that of the unit vector $\mathbf{u}_i$, the direction of the reflected plane wave $\mathbf{u}_r$, and the normal to the surface $\mathbf{n}$, then the law of reflection is expressed by the relation between these unit vectors

$$\mathbf{u}_r = \mathbf{u}_i - 2(\mathbf{n} \,.\, \mathbf{u}_i)\mathbf{n} \quad . \tag{6.1}$$

A plane wave of arbitrary polarization, incident in the direction $\mathbf{u}_i$, can be written as the sum

$$\mathbf{e}_i = \mathbf{e}_i^{\,+} + \mathbf{e}_i^{\,-} \tag{6.2}$$

where the amplitude and phase relationships between $\mathbf{e}_i^+$ and $\mathbf{e}_i^-$ determine whether the incident plane wave is linearly, circularly or elliptically polarized (see Appendix A.5). It is then clear from Fig. 6.1 that, in order to satisfy the boundary conditions, the normal component $(\mathbf{n}.\mathbf{e}_i)\mathbf{n}$ is unchanged on reflection, while the tangential component $\mathbf{e}_i - (\mathbf{n} \cdot \mathbf{e}_i)\mathbf{n}$ is reversed in sign. Hence

$$\mathbf{e}_r = -\mathbf{e}_i + 2(\mathbf{n} \cdot \mathbf{e}_i)\mathbf{n} \tag{6.3}$$

specifies the relationship between $\mathbf{e}_r$ and $\mathbf{e}_i$, the reflected and incident electric fields respectively, at the surface.

Equations (6.1) and (6.3) provide the basic solution to the problem of a plane wave incident on a conducting plane surface. They will now be used to find the magnetic field at the surface, given $\mathbf{e}_i$ and $\mathbf{u}_i$ (which is a result that will be needed in Chapter 8) and the corresponding surface current density.

The point O in the plane depicted in Fig. 6.2 is both a geometrical and a zero-phase reference. If the electric field vector of the incident plane wave is $\mathbf{e}_i$,

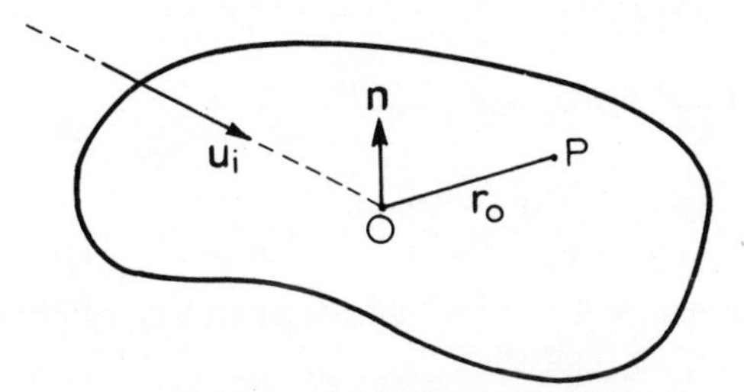

Fig. 6.2 – A point P in the plane.

and its direction is $\mathbf{u}_i$, then the incident electromagnetic field at a point in the medium above the surface is

$$\mathbf{E}_i(\mathbf{r}) = \mathbf{e}_i \exp(-jk\mathbf{u}_i \cdot \mathbf{r}) \tag{6.4}$$

and

$$\mathbf{H}_i(\mathbf{r}) = Y\mathbf{u}_i \times \mathbf{E}_i(\mathbf{r}) = Y\mathbf{u}_i \times \mathbf{e}_i \exp(-jk\mathbf{u}_i \cdot \mathbf{r}) \tag{6.5}$$

where Y is the characteristic plane-wave admittance of the medium. The reflected electromagnetic field is

$$\mathbf{E}_r(\mathbf{r}) = \mathbf{e}_r \exp(-jk\mathbf{u}_r \cdot \mathbf{r}) \tag{6.7}$$

and

$$\mathbf{H}_r(\mathbf{r}) = Y\mathbf{u}_r \times \mathbf{e}_r \exp(-jk\mathbf{u}_r \cdot \mathbf{r}) \tag{6.8}$$

with equations (6.1) and (6.3) substituted. The total electric and magnetic fields are

$$\mathbf{E}(\mathbf{r}) = \mathbf{E}_i(\mathbf{r}) + \mathbf{E}_r(\mathbf{r}) \tag{6.9}$$

and

$$\mathbf{H}(\mathbf{r}) = \mathbf{H}_i(\mathbf{r}) + \mathbf{H}_r(\mathbf{r}) \quad . \tag{6.10}$$

At the surface ($\mathbf{r} = \mathbf{r}_o$), the total electric field is

$$\mathbf{E}(\mathbf{r}_o) = \mathbf{e}_i \exp(-jk\mathbf{u}_i \, . \, \mathbf{r}_o) + \mathbf{e}_r \exp(-jk\mathbf{u}_r \, . \, \mathbf{r}_o) \quad . \tag{6.11}$$

Equation (6.1), together with the condition that $\mathbf{n} \, . \, \mathbf{r}_o = 0$, yields

$$\mathbf{u}_r \, . \, \mathbf{r}_o = \mathbf{u}_i \, . \, \mathbf{r}_o \tag{6.12}$$

and substitution of (6.3) gives

$$\mathbf{E}(\mathbf{r}_o) = 2(\mathbf{n} \, . \, \mathbf{e}_i)\mathbf{n} \exp(-jk\mathbf{u}_i \, . \, \mathbf{r}_o) \tag{6.13}$$

which is wholly normal, and checks that the boundary condition on the total electric field at the surface is satisfied.

The total magnetic field at the surface is

$$\mathbf{H}(\mathbf{r}_o) = Y[\mathbf{u}_i \times \mathbf{e}_i + \mathbf{u}_r \times \mathbf{e}_r] \exp(-jk\mathbf{u}_i \, . \, \mathbf{r}_o) \tag{6.14}$$

which, with (6.1) and (6.3) substituted, yields

$$\mathbf{H}(\mathbf{r}_o) = 2Y[(\mathbf{n} \, . \, \mathbf{e}_i)\mathbf{u}_i \times \mathbf{n} + (\mathbf{n} \, . \, \mathbf{u}_i)\mathbf{n} \times \mathbf{e}_i] \exp(-jk\mathbf{u}_i \, . \, \mathbf{r}_o) \tag{6.15}$$

which is totally tangential to the surface (cf. equation B.2.18 in appendix B). The vector triple product identities of equation (2.13) can be used to rewrite this as

$$\mathbf{H}(\mathbf{r}_o) = 2Y[\mathbf{u}_i \times \mathbf{e}_i - (\mathbf{n} \, . \, \mathbf{u}_i \times \mathbf{e}_i)\mathbf{n}] \exp(-jk\mathbf{u}_i \, . \, \mathbf{r}_o) \quad . \tag{6.16}$$

Physically, this means that the total magnetic field at the surface of an infinite, flat perfect conductor is equal to twice the tangential component of the incident magnetic field. Fig. 6.3 shows this very clearly.

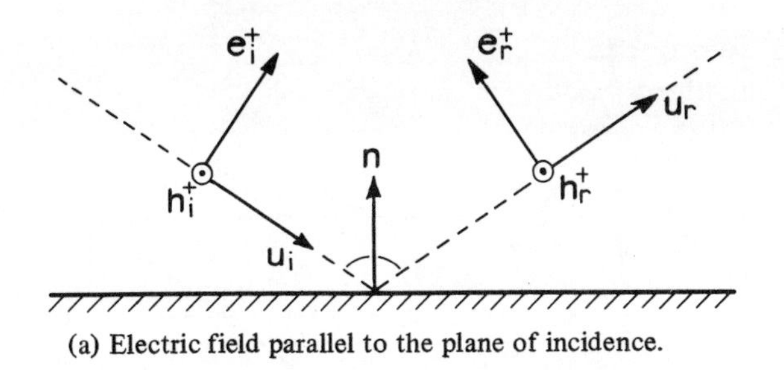

(a) Electric field parallel to the plane of incidence.

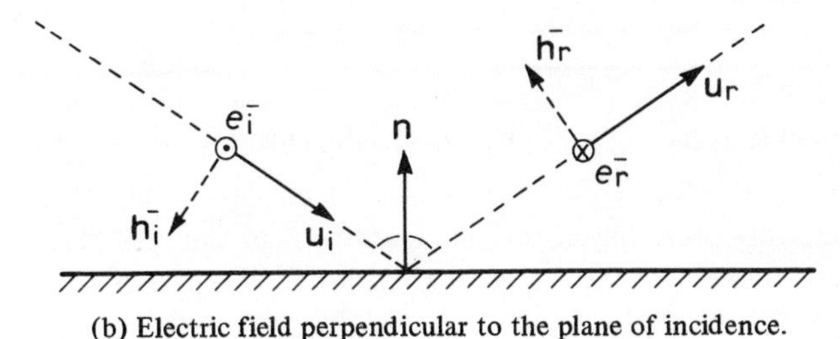

(b) Electric field perpendicular to the plane of incidence.

Fig. 6.3 – Electric and magnetic fields of a plane wave on reflection from a perfectly conducting infinite plane.

The surface current density (see equation B. 2.20) corresponding to the total magnetic field of equation (6.16) is then

$$\mathbf{J}_s(\mathbf{r}_o) = \mathbf{n} \times \mathbf{H}(\mathbf{r}_o) = 2Y\mathbf{n} \times (\mathbf{u}_i \times \mathbf{e}_i) \exp(-jk\mathbf{u}_i \cdot \mathbf{r}_o) \tag{6.18}$$

which has the same amplitude as the (tangential) magnetic field but is directed at right angles to it.

Exercise

Show that a right-handed (clockwise) circularly polarized plane wave becomes left-handed (counterclockwise) circularly polarized on reflection from a perfectly conducting plane. Find the induced surface current density.

6.2 REFLECTION FROM FLAT, FINITE SURFACES

Suppose that the rectangular, flat conducting plate, shown lying in the x-y plane in Fig. 6.4 is illuminated by a plane wave of complex amplitude E_o. Take the direction of incidence to be in the x-z plane, but inclined at angle θ_o to the z-axis. Over the plate ($a \times b$) the total tangential electric field must be zero, to satisfy the boundary condition. Hence the aperture field corresponding to the reflected field can be taken to be the negative of the tangential component of the incident field over $a \times b$, and zero elsewhere in the x-y aperture plane.

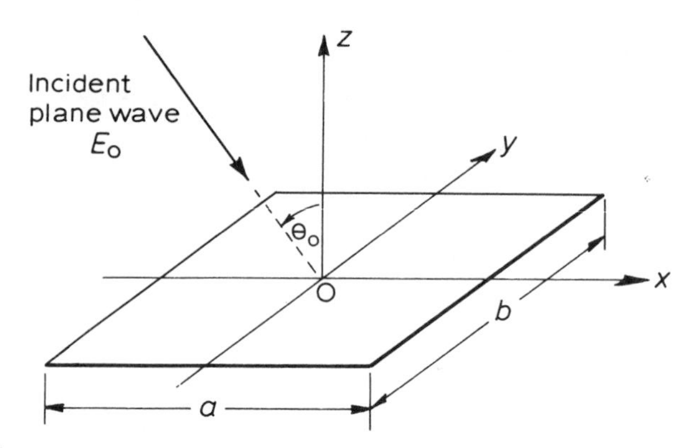

Fig. 6.4 – Rectangular flat plate illuminated by an obliquely incident plane wave.

In effect, this prescription for the aperture field assumes that the field on reflection is the same over the rectangle $a \times b$ as if it were part of an infinite plane. This is the simplest example of what is termed the **tangent-plane approximation of physical optics.** It is justified provided both dimensions a and b are large in comparison with the wavelength, when local diffraction effects are negligible. This can be appreciated from a physical viewpoint by treating each edge as a knife edge. It can then be concluded from the results of section 5.2 that such local diffraction effects are confined to within a fraction of a wavelength of the edge. Also, finite-energy considerations imply that if the fields are singular at the edge then they will at least be integrable singularities.

If the incident wave is + polarized and there is no y component of electric field, the aperture-field distribution for the reflected field will therefore be taken as being

$$E_{ax}(x,y) = -E_o \cos\theta_o \operatorname{rect}_a(x) \operatorname{rect}_b(y) \exp(-jk \sin\theta_o x) \tag{6.19}$$

for both a and b large compared to λ. The corresponding angular plane-wave spectrum is, from equation (3.76),

$$F_x(\alpha,\beta) = -\frac{E_o\gamma_o}{\lambda^2}\int_{-a/2}^{a/2}\int_{-b/2}^{b/2} \exp\{jk\,[(\alpha-\alpha_o)x + \beta y]\}\,dx dy \tag{6.20}$$

$$= -\frac{E_o ab\gamma_o}{\lambda^2}\operatorname{sinc}\left(\frac{a(\alpha-\alpha_o)}{\lambda}\right)\operatorname{sinc}\left(\frac{b\beta}{\lambda}\right) \tag{6.21}$$

where $\alpha_o = \sin\theta_o$ and $\gamma_o = \cos\theta_o$. This spectrum, and hence the far-field pattern (equations (3.82) and (3.101))

$$\mathbf{e}(\theta,\phi) = j2\pi\, F_x(\alpha,\beta)\, [\mathbf{u}_\theta \cos\phi - \mathbf{u}_\phi \cos\theta \sin\phi] \quad , \tag{6.22}$$

is a maximum in the direction $\alpha = \alpha_o, \beta = 0$, the specular (or 'mirror') direction.

Since both a and b are large compared to λ, the main lobe of the reflected radiation will be narrowly confined in angle about the specular direction. The angular width between the first zeros of the main lobe, in the two principal planes are therefore

$$\frac{2\lambda}{\gamma_o a} \text{ and } \frac{2\lambda}{b} \quad . \tag{6.23}$$

Consideration of these beam widths and the spectrum function of equation (6.21) leads to the notion of a 'projected aperture'. This is illustrated in Fig. 6.5, which is a side view of the reflecting surface of Fig. 6.4. The projected aperture field is then $-E_o\exp(-j\frac{1}{2}ka\alpha_o)$ over the rectangle $(a\gamma_o \times b)$. Whether the reflected field at the plate, or the projected aperture field, is used to find the angular

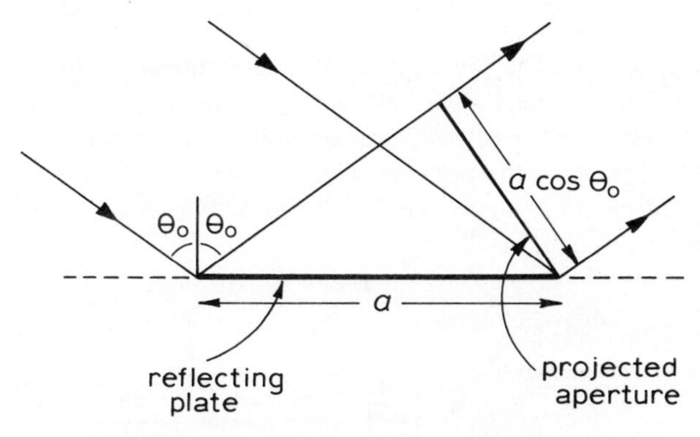

Fig. 6.5 – Projected aperture for a reflecting plate.

spectrum of the reflected field, the results are essentially the same in the angular region of the main lobe and first few side lobes, where agreement with experiment is good (Ross, 1966). Further out in the far sidelobe region of the reflected pattern, edge effects become important and calculations based on the geometrical theory of diffraction (see section 5.2) are more appropriate.

Exercise

Show that the radar backscatter cross-section at normal incidence of a flat conducting plate of area A is $4\pi A^2/\lambda^2$.

Exercise

Deduce the far-field radiation pattern for the reflection of a –polarized plane wave incident on the rectangular flat plate of Fig. 6.4. Compare the E_θ, E_ϕ components with those in the + polarized case.

6.3 REFLECTION FROM CURVED SURFACES – BY AN APPROXIMATE METHOD

In the analysis of reflection from non-planar surfaces by the aperture-field/angular-spectrum method a special difficulty arises. The relationship between the incident and reflected fields can only be specified with any precision at the non-planar surface itself, by applying the boundary conditions on the total tangential fields. (Edges are excluded from this discussion for the moment.) Yet in order to find the angular spectrum without too much mathematical complexity we need to know the aperture field over a plane. This will be achieved in this section in an approximate way be constructing an **equivalent aperture field** using purely physical arguments. The more precise analysis of the next section will then be used to check the validity of this approximate method.

The curved-surface segment in Fig. 6.6, imagined to be part of a larger continuous surface, is assumed to have the profile $z = \zeta(x,y)$, where

$$\zeta(x,y) = -\left(\frac{x^2}{2R_x} + \frac{y^2}{2R_y}\right) \tag{6.24}$$

in which R_x and R_y are the principal radii of curvature at the origin, where the segment is tangential to the x-y plane. This is generally quite a useful description of a curved surface, particularly in antenna applications. If the surface segment has an overall slope, or if the principal radii of curvature do not coincide with the x- and y- directions, allowance can be made for this by reorienting the coordinate axes.

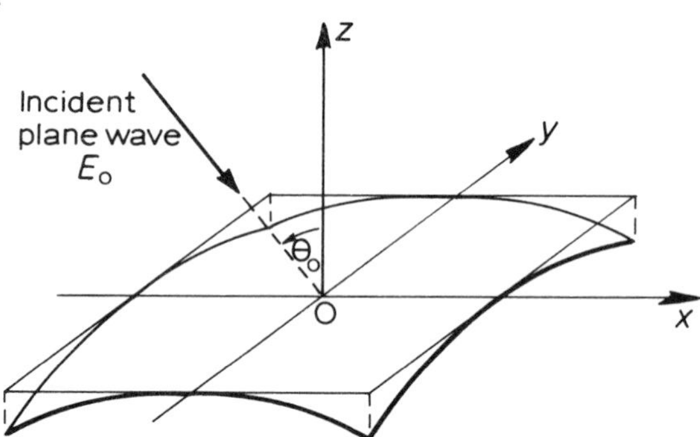

Fig. 6.6 – A curved surface segment, tangential to the x-y plane at the origin.

If a +polarized wave is incident, in a direction coincident with the x-z plane at an angle θ_o to the z-axis, then the approximate equivalent aperture field (in the x-y plane) is

$$E_{ax}(x,y) = -E_o\gamma_o \operatorname{rect}_a(x) \operatorname{rect}_b(y) \exp(-jk\alpha_o x)$$

$$. \exp\left\{-j2k\gamma_o\left(\frac{x^2}{2R_x} + \frac{y^2}{2R_y}\right)\right\} . \tag{6.25}$$

For a –polarized incident plane wave

$$E_{ay}(x,y) = -E_o \, \text{rect}_a(x) \, \text{rect}_b(y) \exp(-jk\alpha_o x)$$

$$. \exp\left\{-j2k\gamma_o\left(\frac{x^2}{2R_x} + \frac{y^2}{2R_y}\right)\right\}. \tag{6.26}$$

These approximate aperture fields have been established on the assumptions that both R_x and R_y are very large compared to the wavelength, that the angle of incidence θ_o is modest, and that the slope of the surface is everywhere small. Then the local behaviour of the field on reflection can be described reasonably accurately in terms of rays, and the excess phase shift at the point (x,y) is approximately $2k\gamma_o\zeta(x,y)$. The remaining effects of the surface are assumed to be the same as for the flat segment examined in the previous section.

The significance of these aperture fields is that they describe the effect of the reflecting surface segment as a spatial-phase (or wavefront) transformer. The linear phase term in equations (6.25) and (6.26) gives the effect of the inclination of the incident wave to the normal of the segment, and the parabolic terms give the effect of its curvature. The spatial-phase transformer action is immediately apparent from the following.

Example

Consider a plane wave incident normally at the apex of a convex paraboloidal surface. Using the equivalent aperture field of equation (6.25), with $\alpha_o = 0$, $\gamma_o = 1$, and

$$R_x = R_y = 2f \ , \tag{6.27}$$

where f is the focal length of the paraboloid,

$$E_{ax}(x,y) = -E_o \, \text{rect}_a(x) \, \text{rect}_b(y) \exp\left(-jk \, \frac{x^2 + y^2}{2f}\right) . \tag{6.28}$$

This aperture field has the form, to a very good approximation close to the origin, of a spherical wave emanating from the focus of the paraboloid as its virtual source. Thus the incident plane wave has been transformed, locally, into a spherical wave.

This result could be used immediately to give the radar backscatter cross-section, the method being essentially that of geometrical optics. But it is more interesting to use the aperture field of equation (6.28) to examine the diffracted field. (An exact result will be obtained in section 6.4 for this case of an apex-illuminated convex paraboloid.) The angular spectrum for the equivalent aperture field of equation (6.28) is

$$F_x(\alpha,\beta) = -\frac{E_o}{\lambda^2}\int_{-a/2}^{a/2}\int_{-b/2}^{b/2} \exp\left\{-jk\,\frac{x^2+y^2}{2f}\right\}\exp\{jk\,(\alpha x + \beta y)\}\,dxdy \quad , \tag{6.29}$$

which can be evaluated in terms of Fresnel integrals (section 5.2). Following a familiar argument, if both a and b are large compared to the Fresnel parameter relevant to this situation, namely $\sqrt{(\lambda f)}$, the angular spectrum in the direction of the z-axis is

$$F_x(0,0) = -\frac{E_o}{\lambda^2}\int_{-\infty}^{\infty}\int_{-\infty}^{\infty} \exp\left\{-jk\,\frac{x^2+y^2}{2f}\right\} dxdy$$

$$= jE_o\,\frac{f}{\lambda} \quad , \tag{6.30}$$

using the standard integral of equation (2.96).

The power flow at a far-field point along the z-axis, a distance r from the apex, is (from equation (3.69))

$$S_r(r,0,0) = \frac{\lambda^2}{2Zr^2}\,|F_x(0,0)|^2 = \frac{|E_o|^2 f^2}{2Zr^2} \quad . \tag{6.31}$$

Since the power density of the incident wave is $|E_o|^2/(2Z)$, the radar backscatter cross-section (defined in section 1.4) is

$$\sigma_{bs} = 4\pi r^2 \,.\, \frac{2Z}{|E_o|^2}\cdot\frac{|E_o|^2 f^2}{2Zr^2} = \pi(2f)^2 \quad . \tag{6.32}$$

The main contribution to the far field on axis can be thought of as being just half that coming from the first Fresnel zone (see section 5.5.4), which is a cap at the apex with a circular rim of radius $\sqrt{(\lambda f)}$. The contributions from the remainder of the surface tend to cancel, and so the approximations made in deriving the equivalent aperture fields of equations (6.25) and (6.26) are rather better than they appear at first sight, at least in a situation such as the present one.

Exercise

What is the radar backscatter cross-section at centimetre wavelengths of a metallic sphere whose diameter is 30 m?

Exercise

Under what conditions can a slightly curved plate be considered to have a radar backscatter cross-section approximating to that of a perfectly flat plate?

The following illustration of the spatial-phase transformer action of a curved surface which is concave with respect to the illuminating field will be useful when discussing the aperture-field technique applied to reflector antennas in Chapter 7.

Example

Consider a segment of a concave paraboloidal surface illuminated by a spherical wave centred on the focus. For convenience, position the rectangular curved segment (Fig. 6.7) at, and tangential to, the origin O, with the focus F in the x-z plane at a distance R from the segment and FO inclined at an angle ψ to the (z-axis) normal to the segment.

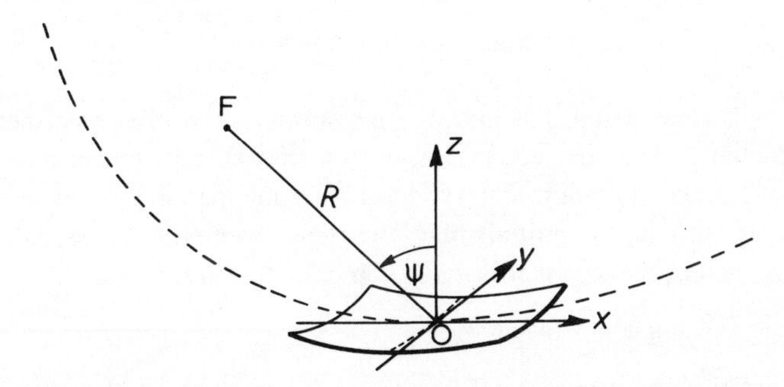

Fig. 6.7 – A segment of a concave paraboloidal surface illuminated from the focus.

The incident field can, in general, be split into the sum of + and – polarized components (cf. equation (6.2)). If the amplitude of the + polarized component is E_0^+ at the surface (the –polarized component can be treated in the same way) the field of the spherical wave incident on the segment at O has the form of a plane wave, obliquely incident at angle ψ, but modified by the parabolic phase term

$$\exp\left\{-jk\left(\frac{x^2\cos^2\psi}{2R}+\frac{y^2}{2R}\right)\right\} \tag{6.33}$$

which is increasingly accurate the smaller the area of the segment.

The profile of the segment is (see section 7.2.1 for the geometrical properties of a paraboloid)

$$\zeta(x,y) = +\frac{x^2}{2R_x}+\frac{y^2}{2R_y} \tag{6.34}$$

with

$$R_x = 2R \sec \psi \quad \text{and} \quad R_y = 2R \cos \psi \ . \tag{6.35}$$

Collecting these items and inserting them into equation (6.25), the equivalent aperture field reflected by the segment is

$$E_{ax}(x,y) = -E_o^+ \cos \psi \ \mathrm{rect}_a(x) \ \mathrm{rect}_b(y)$$

$$\exp(-jkx \cos \psi) \exp\left\{-jk\left(\frac{x^2 \cos^2 \psi}{2R} + \frac{y^2}{2R}\right)\right\}$$

$$\exp\left\{+j2k \cos \psi \left(\frac{x^2 \cos \psi}{4R} + \frac{y^2}{4R \cos \psi}\right)\right\}$$

$$= -E_o^+ \cos \psi \ \mathrm{rect}_a(x) \ \mathrm{rect}_b(y) \exp(-jkx \cos \psi) \tag{6.36}$$

which has a plane wavefront. Of course, this result is just another way of expressing the well known fact, from geometrical optics, that all rays emanating from the focus of a concave paraboloidal reflector become parallel to each other on reflection. It should be noted that the field over the projected aperture (cf. Fig. 6.5) is simply $-E_o^+$, if it is made coincident with O.

Exercise

Use the spatial-phase transformation technique to confirm that, for a hyperboloidal reflector, when the source is at one focal point the reflected field emanates from the other focal point as its virtual source.

6.4 REFLECTION FROM CURVED SURFACES – BY A MORE PRECISE METHOD

Some of the uncertainties inherent in the 'equivalent aperture field' method of treating reflection from non-planar surfaces, described in the previous section, can be circumvented in the following way. Assuming that all parts of the surface are directly illuminated (that is, there is no shadowing of one part of the surface by another part), since the total tangential electric field must be zero at all points on the reflector, the tangential electric field on reflection is known at every point. Each point can then be treated as a radiating Huygens element, and the precise form of Huygens' principle (derived in section 3.5.2) used to derive the total reflected field by integrating over the entire surface. There is a choice to be made between using the general form of Huygens' principle (equation (3.99)) and the far-field form (equation (3.100)). We choose the

Exercise

Under what conditions can a slightly curved plate be considered to have a radar backscatter cross-section approximating to that of a perfectly flat plate?

The following illustration of the spatial-phase transformer action of a curved surface which is concave with respect to the illuminating field will be useful when discussing the aperture-field technique applied to reflector antennas in Chapter 7.

Example

Consider a segment of a concave paraboloidal surface illuminated by a spherical wave centred on the focus. For convenience, position the rectangular curved segment (Fig. 6.7) at, and tangential to, the origin O, with the focus F in the x-z plane at a distance R from the segment and FO inclined at an angle ψ to the (z-axis) normal to the segment.

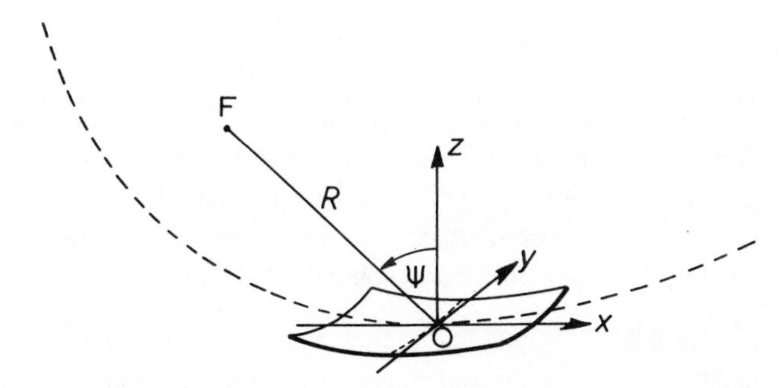

Fig. 6.7 – A segment of a concave paraboloidal surface illuminated from the focus.

The incident field can, in general, be split into the sum of + and – polarized components (cf. equation (6.2)). If the amplitude of the + polarized component is E_0^+ at the surface (the –polarized component can be treated in the same way) the field of the spherical wave incident on the segment at O has the form of a plane wave, obliquely incident at angle ψ, but modified by the parabolic phase term

$$\exp\left\{-jk\left(\frac{x^2\cos^2\psi}{2R}+\frac{y^2}{2R}\right)\right\} \tag{6.33}$$

which is increasingly accurate the smaller the area of the segment.

The profile of the segment is (see section 7.2.1 for the geometrical properties of a paraboloid)

$$\zeta(x,y) = +\frac{x^2}{2R_x}+\frac{y^2}{2R_y} \tag{6.34}$$

with

$$R_x = 2R \sec \psi \quad \text{and} \quad R_y = 2R \cos \psi \ . \tag{6.35}$$

Collecting these items and inserting them into equation (6.25), the equivalent aperture field reflected by the segment is

$$E_{ax}(x,y) = -E_0^+ \cos \psi \ \text{rect}_a(x) \ \text{rect}_b(y)$$

$$\exp(-jkx \cos \psi) \exp\left\{-jk\left(\frac{x^2 \cos^2 \psi}{2R} + \frac{y^2}{2R}\right)\right\}$$

$$\exp\left\{+j2k \cos \psi \left(\frac{x^2 \cos \psi}{4R} + \frac{y^2}{4R \cos \psi}\right)\right\}$$

$$= -E_0^+ \cos \psi \ \text{rect}_a(x) \ \text{rect}_b(y) \exp(-jkx \cos \psi) \tag{6.36}$$

which has a plane wavefront. Of course, this result is just another way of expressing the well known fact, from geometrical optics, that all rays emanating from the focus of a concave paraboloidal reflector become parallel to each other on reflection. It should be noted that the field over the projected aperture (cf. Fig. 6.5) is simply $-E_0^+$, if it is made coincident with O.

Exercise

Use the spatial-phase transformation technique to confirm that, for a hyperboloidal reflector, when the source is at one focal point the reflected field emanates from the other focal point as its virtual source.

6.4 REFLECTION FROM CURVED SURFACES – BY A MORE PRECISE METHOD

Some of the uncertainties inherent in the 'equivalent aperture field' method of treating reflection from non-planar surfaces, described in the previous section, can be circumvented in the following way. Assuming that all parts of the surface are directly illuminated (that is, there is no shadowing of one part of the surface by another part), since the total tangential electric field must be zero at all points on the reflector, the tangential electric field on reflection is known at every point. Each point can then be treated as a radiating Huygens element, and the precise form of Huygens' principle (derived in section 3.5.2) used to derive the total reflected field by integrating over the entire surface. There is a choice to be made between using the general form of Huygens' principle (equation (3.99)) and the far-field form (equation (3.100)). We choose the latter, partly because it is simpler, and we often wish to know only the far field. But there is no loss of information in doing so, since the far field yields the angular spectrum directly, and the angular spectrum describes the field everywhere in the half-space clear of the reflecting structure. The method of reflecting Huygens elements, as it will be called, will first be elaborated. It will then be applied to the finite curved surface already considered, and then to a reflecting surface for which the reflected field is known exactly.

6.4.1 The method of reflecting Huygens elements

In Fig. 6.8 the field incident on the reflecting element at O′ is a plane wave $\mathbf{e}_o$

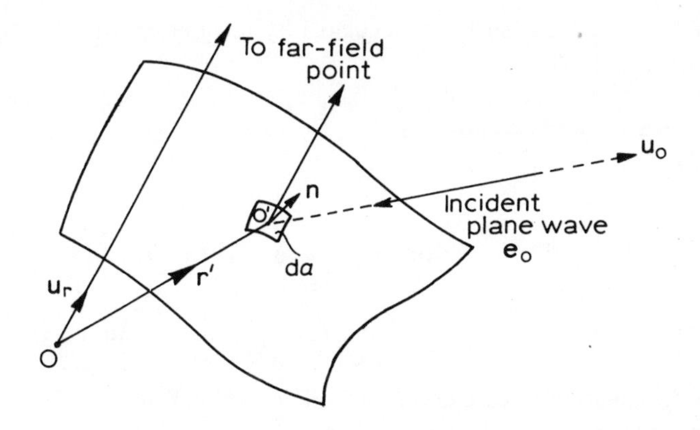

Fig. 6.8 – Element da on a reflecting surface.

incident from the direction $\mathbf{u}_o$. The geometrical origin O is also the phase-reference point, and so with the element a distance $\mathbf{r}'$ from O the field incident on the element is

$$\mathbf{e}_i = \mathbf{e}_o \exp\{jk\, \mathbf{u}_o \cdot \mathbf{r}'\} \tag{6.37}$$

The reflected field, at the element, is then, from equation (6.3),

$$\mathbf{e}_r = [-\mathbf{e}_o + 2\,(\mathbf{n} \cdot \mathbf{e}_o)\mathbf{n}] \exp\{jk\, \mathbf{u}_o \cdot \mathbf{r}'\} \tag{6.38}$$

where $\mathbf{n}$ is the surface element normal. This ensures that the total tangential electric field at the element is zero.

Noting that the second term in the square brackets of equation (6.38) is normal to the element, and so does not contribute, the field at the element can be taken as

$$\mathbf{E}_a = -\,\mathbf{e}_o \exp\{jk\, \mathbf{u}_o \cdot \mathbf{r}'\}\ . \tag{6.39}$$

The elemental contribution to the far field is then given by equation (3.100) as

$$\underset{P\to\infty}{\mathrm{d}\mathbf{E}(P)} = j\,\frac{2\pi}{\lambda^2}\,\frac{\exp(-jkr)}{kr}\,\exp\{jk(\mathbf{u}_r\,.\,\mathbf{r}' + \mathbf{u}_o\,.\,\mathbf{r}')\}\,\mathbf{u}_r\times(\mathbf{n}\times\mathbf{e}_o)\,da$$

$$= j\,\frac{2\pi}{\lambda^2}\,\frac{\exp(-jkr)}{kr}\,\exp\{jk(\mathbf{u}_r + \mathbf{u}_o)\,.\,\mathbf{r}'\}$$

$$[(\mathbf{u}_r\,.\,\mathbf{e}_o)\mathbf{n} - (\mathbf{u}_r\,.\,\mathbf{n})\mathbf{e}_o]\,da \quad . \qquad (6.40)$$

The total far field can then be calculated by integrating over the surface S of the reflector, giving

$$\underset{P\to\infty}{\mathbf{E}(P)} = j\,2\pi\,\frac{\exp(-jkr)}{kr}\,.\,\frac{1}{\lambda^2}\int_S \exp\{jk(\mathbf{u}_r + \mathbf{u}_o)\,.\,\mathbf{r}'\}$$

$$[(\mathbf{u}_r\,.\,\mathbf{e}_o)\mathbf{n} - (\mathbf{u}_r\,.\,\mathbf{n})\mathbf{e}_o]\,da \quad . \qquad (6.41)$$

In general $\mathbf{e}_o$ and $\mathbf{u}_o$, as well as $\mathbf{r}'$ and $\mathbf{n}$, vary with position on the surface of the reflector.

The formulation in equation (6.41) of reflection from a conducting surface makes no allowance for shadowing. It also takes no account of reradiation by one part of the surface of the radiation from another part, and so must be restricted in application to continuous surfaces whose radius of curvature is everywhere reasonably large compared to the wavelength. It is also inaccurate in the vicinity of edges, and so is further restricted to reflecting surfaces whose overall dimensions are large compared to the wavelength. A final restriction is that the illuminating field must have the form, at least locally and at all points of the surface, of a plane wave. However, none of these limitations is too restrictive in applications to microwave reflector antennas.

In the Cartesian coordinate system of Fig. 6.9 the reflecting Huygens element formula of equation (6.40) acquires the following form. The reflector surface has the general profile

$$z = \zeta(x,y) \qquad (6.42)$$

which is assumed to be single-valued, and to possess finite first and second derivatives such that the physical limitations described in the previous paragraph are met. The normal at each point of the surface,

$$\mathbf{n} = \mathbf{u}_x\,n_x + \mathbf{u}_y\,n_y + \mathbf{u}_z\,n_z \qquad (6.43)$$

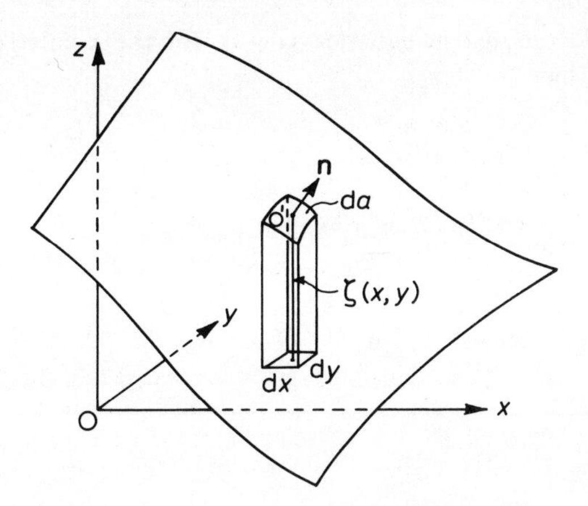

Fig. 6.9 – The reflecting surface of Fig. 6.8, in Cartesian coordinates.

is given by

$$\mathbf{n} = \frac{\nabla\phi}{\|\nabla\phi\|} \quad , \text{ with } \phi(x,y,z) = z - \zeta(x,y) \quad , \tag{6.44}$$

where $\|\nabla\phi\|$, is the norm of the vector gradient $\nabla\phi$. The vector position of the element at O′ is

$$\mathbf{r}' = \mathbf{u}_x x + \mathbf{u}_y y + \mathbf{u}_z \zeta(x,y) \tag{6.45}$$

The incident field is conveniently split into the sum of x- and y- polarized plane waves, such that

$$\begin{aligned}\mathbf{e}_o &= e_{ox}\,(\mathbf{u}_x - \frac{\alpha_o}{\gamma_o}\,\mathbf{u}_z) + e_{oy}\,(\mathbf{u}_y - \frac{\beta_o}{\gamma_o}\,\mathbf{u}_z)\\ &= \mathbf{e}_{ox} + \mathbf{e}_{oy}\end{aligned} \tag{6.46}$$

arriving from the direction

$$\mathbf{u}_o = \mathbf{u}_x\,\alpha_o + \mathbf{u}_y\,\beta_o + \mathbf{u}_z\,\gamma_o \tag{6.47}$$

The area of the reflecting element is

$$da = \frac{\mathrm{d}x\,\mathrm{d}y}{\mathbf{n}\,.\,\mathbf{u}_z} = \frac{\mathrm{d}x\,\mathrm{d}y}{n_z} \tag{6.48}$$

The vector triple product in equation (6.40), for the x-polarized part of the incident field, is then

$$\mathbf{u}_r \times (\mathbf{n} \times \mathbf{e}_{ox}) = (\mathbf{u}_r \cdot \mathbf{e}_{ox})\mathbf{n} - (\mathbf{u}_r \cdot \mathbf{n})\mathbf{e}_{ox}$$

$$= (\alpha - \frac{\alpha_o}{\gamma_o}\gamma)\ [\mathbf{u}_x n_x + \mathbf{u}_y n_y + \mathbf{u}_z n_z]\, e_{ox}$$

$$- (\alpha n_x + \beta n_y + \gamma n_z)[\mathbf{u}_x - \frac{\alpha_o}{\gamma_o}\mathbf{u}_z]\ e_{ox}$$

and

$$\frac{\mathbf{u}_r \times (\mathbf{n} \times \mathbf{e}_{ox})}{n_z} = e_{ox}\left[-\mathbf{u}_x\left(\frac{n_x}{n_z}\frac{\alpha_o}{\gamma_o}\gamma + \beta\frac{n_y}{n_z} + \gamma\right)\right.$$

$$+ \mathbf{u}_y\left(\frac{n_y}{n_z}\alpha - \frac{n_y}{n_z}\frac{\alpha_o}{\gamma_o}\gamma\right)$$

$$\left. + \mathbf{u}_z\left(\frac{n_x}{n_z}\frac{\alpha_o}{\gamma_o}\alpha + \frac{n_y}{n_z}\frac{\alpha_o}{\gamma_o}\beta + \alpha\right)\right]\ . \tag{6.49}$$

It is interesting to note, even at this stage, that this elemental reflected field, which arises from a purely x-polarized incident wave, is the sum of an x-polarized and a y-polarized plane wave in the far field. But if the reflector is everywhere parallel to the x-y plane, and $n_x = n_y = 0$, the reflected field reduces to being x-polarized.

Exercise

Confirm that the field of equation (6.49) can be split into the following sum of an x-polarized and a y-polarized plane wave:

$$\frac{\mathbf{u}_r \times (\mathbf{n} \times \mathbf{e}_{ox})}{n_z} = -e_{ox}\left(1 + \frac{n_x}{n_z}\frac{\alpha_o}{\gamma_o} + \frac{n_y}{n_z}\frac{\beta}{\gamma}\right)(\mathbf{u}_x\gamma - \mathbf{u}_z\alpha)$$

$$+ e_{ox}\frac{n_y}{n_z}\left(\frac{\alpha}{\gamma} - \frac{\alpha_o}{\gamma_o}\right)(\mathbf{u}_y\gamma - \mathbf{u}_z\beta) \tag{6.49a}$$

Hence determine all the conditions under which there is no y-polarized component in the reflected field.

For the y-polarized part of the incident field, again expanding the vector triple product, we have that

$$\frac{\mathbf{u}_r \times (\mathbf{n} \times \mathbf{e}_{oy})}{n_z} = e_{oy}\left[\mathbf{u}_x \frac{n_x}{n_z}\left(\beta - \frac{\beta_o}{\gamma_o}\right)\right.$$

$$- \mathbf{u}_y\left(\frac{n_x}{n_z}\alpha + \frac{n_y}{n_z}\frac{\beta_o}{\gamma_o}\gamma + \gamma\right)$$

$$\left. + \mathbf{u}_z\left(\frac{n_x}{n_z}\frac{\beta_o}{\gamma_o}\alpha + \frac{n_y}{n_z}\frac{\beta_o}{\gamma_o}\beta + \beta\right)\right] \tag{6.50}$$

$$= e_{oy}\frac{n_x}{n_z}\left(\frac{\beta}{\gamma} - \frac{\beta_o}{\gamma_o}\right)(\mathbf{u}_x\gamma - \mathbf{u}_z\alpha)$$

$$- e_{oy}\left(1 + \frac{n_x}{n_z}\frac{\alpha}{\gamma} + \frac{n_y}{n_z}\frac{\beta_o}{\gamma_o}\right)(\mathbf{u}_y\gamma - \mathbf{u}_z\beta) \tag{6.50a}$$

which again consists of the sum of x- and y-polarized plane waves. In the following it is understood that

$$\frac{\mathbf{u}_r \times (\mathbf{n} \times \mathbf{e}_o)}{n_z} = \frac{\mathbf{u}_r \times (\mathbf{n} \times \mathbf{e}_{ox})}{n_z} + \frac{\mathbf{u}_r \times (\mathbf{n} \times \mathbf{e}_{oy})}{n_z} \tag{6.51}$$

with equations (6.49) and (6.50) substituted.

The phase term in equation (6.40) is

$$k(\mathbf{u}_r + \mathbf{u}_o) \cdot \mathbf{r}' = k\ [(\alpha_o + \alpha)\,x + (\beta_o + \beta)y + (\gamma_o + \gamma)\zeta\,(x,y)] \quad . \tag{6.52}$$

Gathering the preceding expressions together and substituting them into equation (6.40), the elemental reflected far field is

$$\underset{P\to\infty}{\mathrm{d}\mathbf{E}(P)} = j\ \frac{2\pi}{\lambda^2}\ \frac{\exp(-jkr)}{kr}\ \frac{\mathbf{u}_r \times (\mathbf{n} \times \mathbf{e}_o)}{n_z}$$

$$\exp\{jk\ [(\alpha_o + \alpha)x + (\beta_o + \beta)y + (\gamma_o + \gamma)\zeta\,(x,y)]\}\ \mathrm{d}x\ \mathrm{d}y \quad . \tag{6.53}$$

This expression can then be integrated over the vertical projection of the surface on to the x-y plane to give the total reflected far field. Thus

$$\underset{P\to\infty}{\mathbf{E}(P)} = j2\pi \frac{\exp(-jkr)}{kr} \frac{1}{\lambda^2} \int_{-\infty}^{\infty}\int_{-\infty}^{\infty} \frac{\mathbf{u}_r \times (\mathbf{n} \times \mathbf{e}_o)}{n_z}$$

$$\exp\{jk\,[\alpha_o x + \beta_o y + (\gamma_o + \gamma)\zeta(x,y)]\}$$

$$\exp\{jk\,(\alpha x + \beta y)\}\,\mathrm{d}x\mathrm{d}y \quad . \tag{6.54}$$

Comparing this with the general far-field equation (3.80), having substituted the details for the vector triple product (equation (6.51)), it can be seen, using equations (3.76) and (3.77), that a more precise version of the equivalent aperture field proposed in the previous section is

$$\mathbf{E}_a(x,y) = \mathbf{u}_x\,E_{ax}(x,y) + \mathbf{u}_y\,E_{ay}(x,y) \quad , \tag{6.55}$$

where

$$E_{ax}(x,y) = -\left[e_{ox}\left(1 + \frac{n_x}{n_z}\frac{\alpha_o}{\gamma_o} + \frac{n_y}{n_z}\frac{\beta}{\gamma}\right) + e_{oy}\frac{n_x}{n_z}\left(\frac{\beta_o}{\gamma_o} - \frac{\beta}{\gamma}\right)\right]$$

$$\exp\{jk\,[\alpha_o x + \beta_o y + (\gamma_o + \gamma)\zeta(x,y)]\} \tag{6.56}$$

and

$$E_{ay}(x,y) = -\left[e_{ox}\frac{n_y}{n_z}\left(\frac{\alpha_o}{\gamma_o} - \frac{\alpha}{\gamma}\right) + e_{oy}\left(1 + \frac{n_x}{n_z}\frac{\alpha}{\gamma} + \frac{n_y}{n_z}\frac{\beta_o}{\gamma_o}\right)\right]$$

$$\exp\{jk\,[\alpha_o x + \beta_o y + (\gamma_o + \gamma)\zeta(x,y)]\} \tag{6.57}$$

This more precise equivalent aperture field is fairly complex, in that it contains cross-polarized terms and depends on the local surface slope and angle of observation, as well as on the nature and angle of incidence of the incident field. But the aperture field of equations (6.55) to (6.57) will prove to be very useful in matters both theoretical and practical. An immediate use for it will be to check the validity of the equivalent aperture field of section 6.3, which was based on a physical argument. It will be used repeatedly in Chapter 7, in the analysis of paraboloidal reflector antannas, whenever a more precise description of the behaviour of the fields is required. The more precise equivalent aperture field also has a much wider application in scattering theory and radar, where it could be used to describe and understand the nature of the fields scattered by irregularly shaped conducting bodies and boundaries. (Earlier, more approximate, versions can be found in Rice (1951) and Clarke and Hendry (1964).)

6.4.2 Application to a doubly curved surface

We return now to the example considered in section 6.3, where the reflecting surface was doubly parabolic, convex and tangential to the origin, with two principal radii of curvature (R_x, R_y) and illuminated by an x-polarized plane wave of amplitude E_o incident from the direction ($-\alpha_o$, 0). For the surface profile described by equation (6.24), the ratios of its normal components (see equations (6.43) and (6.44)) are

$$\frac{n_x}{n_z} = \frac{x}{R_x} \text{ and } \frac{n_y}{n_z} = \frac{y}{R_y} \tag{6.58}$$

Hence equations (6.55) to (6.57) give the aperture field, calculated by the method of reflecting Huygens elements, as

$$\mathbf{E}_a(x,y) = -E_o\,\gamma_o\,\text{rect}_a(x)\,\text{rect}_b(y)\,\exp\,(-jk\,\alpha_o x)$$

$$\exp\left\{-jk\,(\gamma_o + \gamma)\left(\frac{x^2}{2R_x} + \frac{y^2}{2R_y}\right)\right\}$$

$$\left[\mathbf{u}_x\left(1 - \frac{x}{R_x}\frac{\alpha_o}{\gamma_o} + \frac{y}{R_y}\frac{\beta}{\gamma}\right) - \mathbf{u}_y\,\frac{y}{R_y}\left(\frac{\alpha_o}{\gamma_o} + \frac{\alpha}{\gamma}\right)\right], \tag{6.59}$$

in which a and b are the sides of the rectangle which the vertical projection of the surface on the x-y plane is presumed to have. If a and b are each many wavelengths long the principal direction in which the reflected field travels will be (α_o,0) for which $\gamma = \gamma_o$. (In the backscatter direction, which is also often of interest, $\gamma = \gamma_o$ again). Substituting this value in the phase term, and noting that the maximum values of x and y are respectively $a/2$ and $b/2$, then provided $a \ll R_x$ and $b \ll R_y$, or in other words the surface slope is everywhere small, the equivalent aperture field of equation (6.25) is a good approximation.

6.4.3 An apex-illuminated convex paraboloidal reflector

Schensted (1955) has given an exact solution of the problem of scattering by a perfectly conducting, convex paraboloid (not truncated), illuminated by a plane electromagnetic wave incident on its axis. We will use the method of reflecting Huygens elements to find the reflected field radiating into the half-space $z > 0$. The profile of the convex paraboloid is given by

$$\zeta(x,y) = -\frac{x^2 + y^2}{4f} \tag{6.60}$$

where f is the focal length. The aperture field for an x-polarized plane wave will therefore be that of equation (6.59) with $R_x = R_y = 2f$, $\alpha_o = 0$, and with a and b extending to infinity. Thus

$$\mathbf{E}_a(x,y) = -E_o \exp\left\{-jk(1+\gamma)\frac{x^2+y^2}{4f}\right\}$$

$$\left[\mathbf{u}_x\left(1+\frac{y}{2f}\frac{\beta}{\gamma}\right) - \mathbf{u}_y\frac{y}{2f}\frac{\alpha}{\gamma}\right], \tag{6.61}$$

and the angular spectra corresponding to the x- and y-polarized parts of the aperture field are respectively

$$F_x(\alpha,\beta) = -\frac{E_o}{\lambda^2}\int_{-\infty}^{\infty}\int_{-\infty}^{\infty}\left(1+\frac{y}{2f}\frac{\beta}{\gamma}\right)\exp\left\{-jk(1+\gamma)\frac{x^2+y^2}{4f}\right\}$$

$$\exp\{jk(\alpha x+\beta y)\}\,dxdy \tag{6.62}$$

and

$$F_y(\alpha,\beta) = \frac{E_o}{\lambda^2}\int_{-\infty}^{\infty}\int_{-\infty}^{\infty}\frac{y}{2f}\frac{\alpha}{\gamma}\exp\left\{-jk(1+\gamma)\frac{x^2+y^2}{4f}\right\}$$

$$\exp\{jk(\alpha x+\beta y)\}\,dxdy\ . \tag{6.63}$$

All these integrals can be evaluated using the standard integral (see Gradshteyn and Rhyzik (1965)).

$$I = \int_{-\infty}^{\infty} e^{-a^2x^2}\,e^{jbx}\,dx = \frac{\sqrt{\pi}}{a}\exp\left\{-\frac{b^2}{4a^2}\right\} \quad (\text{for Re}\ a > 0) \tag{6.64}$$

and its derivative with respect to b,

$$\frac{dI}{db} = j\int_{-\infty}^{\infty} x\,e^{-a^2x^2}\,e^{jbx}\,dx = -\frac{b}{2a^2}\frac{\sqrt{\pi}}{a}\exp\left\{-\frac{b^2}{4a^2}\right\} \tag{6.65}$$

Hence

$$F_x(\alpha,\beta) = jE_o\frac{2f}{\lambda}\frac{1}{(1+\gamma)}\left[1+\frac{\beta^2}{\gamma(1+\gamma)}\right]\exp\{jkf(1-\gamma)\} \tag{6.66}$$

and

$$F_y(\alpha,\beta) = -jE_o \frac{2f}{\lambda} \frac{1}{(1+\gamma)} \frac{\alpha\beta}{\gamma(1+\gamma)} \exp\{jkf(1-\gamma)\} \quad . \tag{6.67}$$

In spherical-polar form, the far field is given by equation (3.82) as

$$\underset{r\to\infty}{\mathbf{E}(r,\theta,\phi)} = j2\pi \frac{\exp(-jkr)}{kr} \Big[\mathbf{u}_\theta \{F_x(\alpha,\beta)\cos\phi + F_y(\alpha,\beta)\sin\phi)\}$$

$$+ \mathbf{u}_\phi \{-F_x(\alpha,\beta)\cos\theta\sin\phi + F_y(\alpha,\beta)\cos\theta\cos\phi\} \Big] \tag{6.68}$$

which becomes, with equations (6.66) and (6.67) substituted.

$$\underset{r\to\infty}{\mathbf{E}(r,\theta,\phi)} = -\frac{\exp(-jkr)}{r} \frac{2f}{1+\cos\theta} \exp\{jkf(1-\cos\theta)\}$$

$$[\mathbf{u}_\theta \cos\phi - \mathbf{u}_\phi \sin\phi] \quad . \tag{6.69}$$

This is precisely the result obtained by Schensted, which is claimed to be exact. That this is so can be seen from the above development, in that the plane waves of equations (6.62) and (6.63) certainly satisfy Maxwell's equations, the boundary condition is satisfied at every point on the surface, and equation (6.69) has the required behaviour at infinity for the solution to be unique.

The above agreement lends support for the validity of the method of reflecting Huygens elements.

Exercise

Show that the amplitude and phase of the radiated field of equation (6.69) can be obtained from geometrical optics.

CHAPTER 7

Analysis of some aperture antennas

The intention in this chapter is to give some examples of the application to antennas of the various aspects of diffraction we have considered in previous chapters. The antennas will be of the type known as aperture antennas, used in the microwave range of the radio-frequency spectrum. Roughly speaking, we will follow the list of examples given in the introductory chapter. But the emphasis will be on the techniques of analysis, and on the broader conclusions that can be drawn, rather than on obtaining detailed results. The latter depend very much on the particular purpose for which an antenna is intended, and usually involves the extensive use of numerical techniques. We hope to provide a basic analytical understanding, which should lead ultimately to better design and a more efficient use of the powerful numerical techniques that are now available.†

7.1 ELECTROMAGNETIC HORNS

The flared extension of an open-ended waveguide is often used as the primary source of radiation in the design of large, high-gain microwave antennas. The modest gain of the horn means that it is not often used as an antenna in its own right, since every decibel increase in antenna gain can be offset by a corresponding reduction in transmitter power. But the horn is very useful as a gain standard, being robust in construction and being amenable to quite precise prediction.

7.1.1 Pyramidal Horn

Suppose that a rectangular waveguide supports only the dominant TE_{01} mode (see Appendix A. 10). If the waveguide is connected, as shown in Fig. 7.1, to a planar, doubly flared metallic section the result is a pyramidal horn. If the

†Two recent collections of papers (Love (1976 and 1978)) and two review articles (Clarricoats and Poulton (1977) and Rudge and Adatia (1978)) were invaluable aids in the preparation of this chapter.

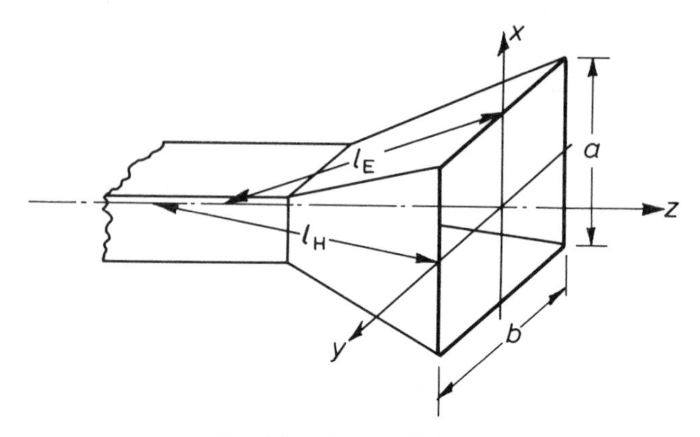

Fig. 7.1 – A pyramidal horn.

flare angle is modest it may be presumed (Schelkunoff, 1943) that the tangential electric field over the exit aperture of the horn is of the same form as the amplitude field distribution of the rectangular TE_{01} mode, with a doubly parabolic phase variation superimposed, corresponding to the flare lengths l_E and l_H in the E and H planes respectively. This may be taken as the aperture field distribution from which the radiation from the horn can be calculated. Thus

$$E_{ax}(x,y) = E_o \operatorname{rect}_{a,b}(x,y) \cos\left(\frac{\pi y}{b}\right) \exp\left\{-\frac{jk}{2}\left(\frac{x^2}{l_E} + \frac{y^2}{l_H}\right)\right\} \tag{7.1}$$

which is purely x-polarized, and whose angular spectrum describing the radiated field is (from equation 3.76)

$$F_x(\alpha,\beta) = \frac{E_o}{2\lambda^2}\int_{-a/2}^{a/2}\int_{-b/2}^{b/2}\left[\exp\left(\frac{j\pi y}{b}\right) + \exp\left(-j\frac{\pi y}{b}\right)\right]$$

$$\exp\left\{-\frac{jk}{2}\left(\frac{x^2}{l_E} + \frac{y^2}{l_H}\right)\right\} \exp\{jk(\alpha x + \beta y)\}\, dx dy \quad . \tag{7.2}$$

Both x- and y- integrations, after completing squares, reduce to Fresnel integrals (defined in section 5.2). The full expression will not be given, but can be readily derived by noting that

$$\int_{-a/2}^{a/2} \exp\left\{-j\frac{kx^2}{2l_E}\right\} \exp(jk\alpha x)\, dx$$

$$= \sqrt{\frac{\lambda l_E}{2}} \exp(j\pi\alpha^2 l_E/\lambda)$$

$$\left[\mathscr{F}\left\{\sqrt{\frac{2}{\lambda l_E}}\left(\frac{a}{2}-\frac{2\alpha}{\lambda}\sqrt{\frac{\lambda l_E}{2}}\right)\right\}+\mathscr{F}\left\{\sqrt{\frac{2}{\lambda l_E}}\left(\frac{a}{2}+\frac{2\alpha}{\lambda}\sqrt{\frac{\lambda l_E}{2}}\right)\right\}\right] \quad (7.3)$$

The y-integration is performed by replacing a by b, l_E by l_H, and $k\alpha$ by $k\beta \pm \pi/b$ in equation (7.3). The far-field radiation is given by equations (3.82) and (3.83), which applies when the Rayleigh criterion (equation (3.84)) is satisfied.

The physical significance of the above can best be seen by returning to the aperture field of equation (7.1). It consists of a uniform amplitude distribution across the a-dimension, a half-cosine tapered amplitude distribution across the b-dimension, and a parabolic phase variation across both. In the special circumstances that the horn is very long (or from a practical standpoint, echoing the Rayleigh criterion, if $l_E > 2a^2/\lambda$ and $l_H > 2b^2/\lambda$) the aperture-field can be regarded as having constant phase. Then the angular spectrum of equation (7.2) reduces to that of equation (3.31) which is plotted in Fig. 3.5. (Note that the beamwidths in the two principal planes will be approximately the same if $b = 1.5a$, but that the sidelobe level in the E-plane is always higher than in the H-plane because of lack of tapering). In practice such a long horn can rarely be used and the effect of the parabolic phase must be accepted.

Figs. 7.2 and 7.3 show the effect, in comparison with the constant-phase (dashed) far-field pattern, of a parabolic phase variation whose maximum

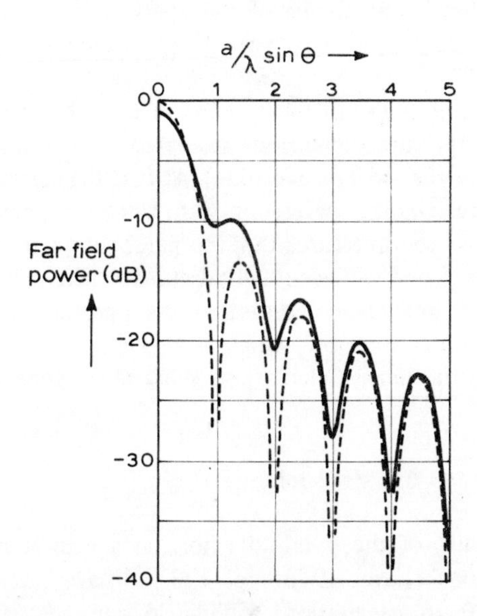

Fig. 7.2 – Far-field power pattern in the E-plane (β=0) of a pyramidal horn with $l_E = a^2/(2\lambda)$ and $l_H = \infty$ (that is, an E-plane sectoral horn). The case of $l_E = l_H = \infty$ is shown dashed.

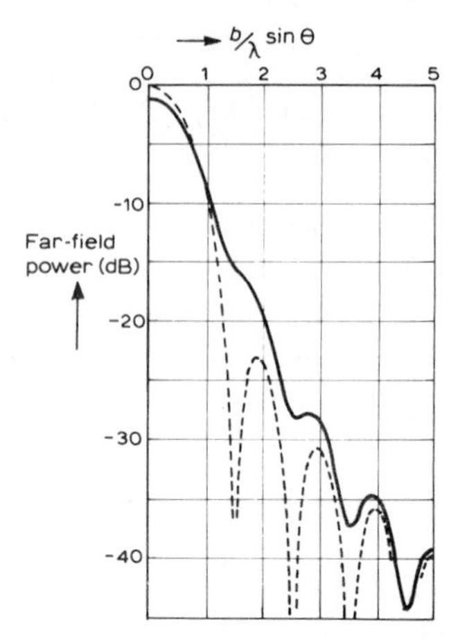

Fig. 7.3 – Far-field power pattern in the H-plane (α=0) of a pyramidal horn with $l_H = b^2/2\lambda$ and $l_E = \infty$ (that is, an H-plane sectoral horn). The case of $l_E = l_H = \infty$ is shown dashed.

phase difference between the centre and all four edges of the aperture is $\pi/2$. It can be seen that the effect is to reduce the on-axis gain slightly, to raise the sidelobe level and fill in the nulls. For some purposes, such as when a pyramidal horn is used as a primary radiator for a reflector antenna, there are some advantages in this result. For one thing, the pattern is rather smoother, though broader, and for another the independent choice of the aperture dimensions and horn lengths gives some flexibility in feed pattern design, which will be seen to be desirable.

From another physical point of view, suppose that the flare lengths are held constant, and the aperture dimensions increased from a minimum value (corresponding to those of the feeder waveguide). At first the increase in aperture area will produce a corresponding increase in gain, since the phase will still be essentially constant across the aperture. But the parabolic phase variation will make itself increasingly felt, and will eventually dominate. The gain will therefore attain a maximum value, and then decrease as the aperture dimensions are further increased. The dimensions for which the gain is a maximum define what is termed the **optimum horn.** Calculations show these dimensions have to be approximately

$$a = \sqrt{(3\lambda l_E)} \text{ and } b = \sqrt{(2\lambda l_H)} \quad . \tag{7.4}$$

With regard to the use of the pyramidal horn as a gain standard, careful experimental investigations (Love, 1976, pp. 128 ff) have revealed that, for gains in the region of 20 dB, calculations of far-field gain (see equation (4.10) based on the evaluation of equation (7.2) in terms of Fresnel integrals are accurate to better than 0.1 dB. At this level of precision the Rayleigh far-field criterion is inadequate. A more precise result is obtained if the slant lengths (l_E and l_H) of the horn are replaced by effective slant lengths

$$l'_E = \frac{rl_E}{r+l_E} \text{ and } l'_H = \frac{rl_H}{r+l_H} \quad , \tag{7.5}$$

where r is the distance at which the gain is required to be known.

For even greater precision the antenna coupling formula must be used, either in its angular-spectrum or aperture-field form.

Exercise

Use the Fresnel-region formulation as in section 5.5.3, to justify the slant length modifications of equation (7.5).

7.1.2 Other types of horns

The straight-sided conical horn has radiation characteristics very similar to those of the pyramidal horn, but the field equations are considerably more complex. Of many ingenious designs, the corrugated horn is important as a widely used primary feed for reflector antennas. It has an axially symmetric, smoothly tapered pattern which leads to low cross-polarization and spillover; its properties are also fairly independent of frequency over a reasonably wide band. The field calculations are very complicated, but the outcome is that in many cases the pattern, which is uniform in ϕ, can be described to quite a good approximation by a Gaussian function of $s = \sin\theta$. The corresponding aperture field is that of equation (3.40).

7.2 PARABOLOID REFLECTOR ANTENNAS

We propose to examine the classical, axisymmetrically front-fed paraboloid reflector antenna in considerable detail. The main reason for this is that the geometry of the structure leads to relatively simple mathematical expressions, so their physical significance is not obscured. Other configurations, such as the horn-paraboloid and offset paraboloid, can be treated using essentially the same methods, as can surfaces other than paraboloids. These will be analysed later but not in such detail.

After an introductory section on the geometry of paraboloidal surfaces the conventional analysis, which obtains the radiation by way of a projected aperture field, will be given. The validity of the projected aperture-field method will then be assessed, both from a physical and a mathematical point of view. The paraboloidal reflector antenna will then be examined as a receiver and the fields in the focal plane derived. Finally the effect of aperture blockage by the presence of the feed and supporting struts, and the effect of reflector-profile irregularities, will be examined.

7.2.1 Geometry of the paraboloid

The paraboloid of focal length f, tangent to the x-y plane at the origin, has the equation

$$\rho^2 = 4fz \quad ; \quad \rho^2 = x^2 + y^2 \tag{7.6}$$

shown in Fig. 7.4.

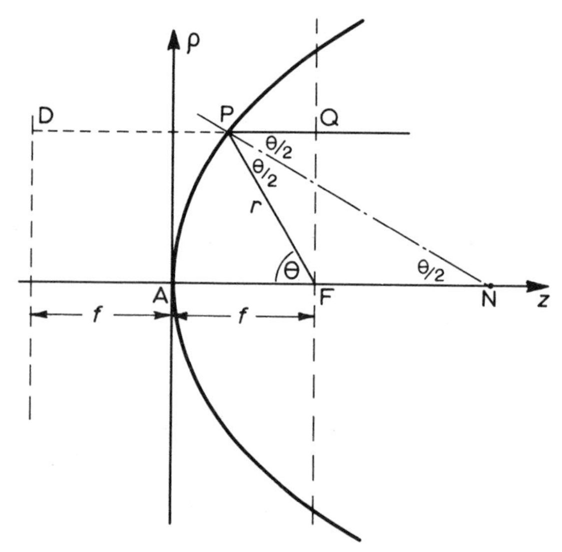

Fig. 7.4 – The paraboloid $\rho^2 = 4fz$.

It is a surface defined such that the distance from the focal point F to any point P on the surface is equal to the perpendicular distance PD of that point from the plane which lies a distance f behind, and is parallel to, the x-y plane. The focal length f is the distance from F to the apex A.

In terms of polar coordinates (r,θ) based on F, it follows from this definition that

$$r = (f - r\cos\theta) + f \tag{7.7}$$

which gives the polar equation of the paraboloid as

$$r = \frac{2f}{1 + \cos\theta} = f\sec^2(\theta/2) \tag{7.8}$$

The 't substitution' $t = \tan(\theta/2)$ is sometimes useful, in terms of which

$$r = f(1 + t^2) \tag{7.9}$$

and

$$\frac{d\theta}{dt} = \frac{2}{1+t^2}, \quad \sin\theta = \frac{2t}{1+t^2}, \quad \cos\theta = \frac{1-t^2}{1+t^2} \quad \text{and} \tan\theta = \frac{2t}{1-t^2} \tag{7.10}$$

The slope of the tangent at P is

$$\frac{d\rho}{dz} = \frac{2f}{\rho} = \frac{1}{t} \tag{7.11}$$

and the slope of the normal PN is therefore

$$-\frac{dz}{d\rho} = -\frac{\rho}{2f} = -t \quad . \tag{7.12}$$

The distance FN is therefore r, and the angle of incidence of the 'ray' FP to the normal is $\theta/2$, so the reflected ray PQ is parallel to the axis. If Q is in the plane parallel to the x-y plane which passes through the focus, the total path length

$$FP + PQ = 2f \tag{7.13}$$

is therefore constant for all rays emanating from F.

The longitudinal radius of curvature of the paraboloid at P is

$$R_l = \left[1 + \left(\frac{dz}{d\rho}\right)^2\right]^{3/2} \Big/ \frac{d^2z}{d\rho^2} = 2\,r \sec(\theta/2) \tag{7.14}$$

which is positive since the surface is always concave when viewed from the z axis. The orthogonal radius of curvature PN is

$$R_o = 2r\cos(\theta/2) \tag{7.15}$$

R_l and R_o are the principal radii of curvature of the paraboloidal surface at P. They both reduce to $2f$ at the apex A.

Exercise

The general polar equation for a conic section is

$$r = f\,\frac{1+e}{1+e\cos\theta} \quad , \tag{7.16}$$

where e is the 'eccentricity'. For a parabola $e = 1$, but for a hyperbola $e > 1$. Find the equation for the surface of a hyperboloid in rectangular coordinates, with the origin of coordinates at the apex and the z-axis of the hyperboloid. Calculate the slopes of the surface and its normal, and the principal radii of curvature, at any point on the surface.

7.2.2 Radiation from an axisymmetrically front-fed paraboloid by the projected aperture-field method

Suppose the paraboloidal reflector of diameter D and focal length f, shown in Fig. 7.5, is fed by a small, x-polarized pyramidal horn placed at its focus.

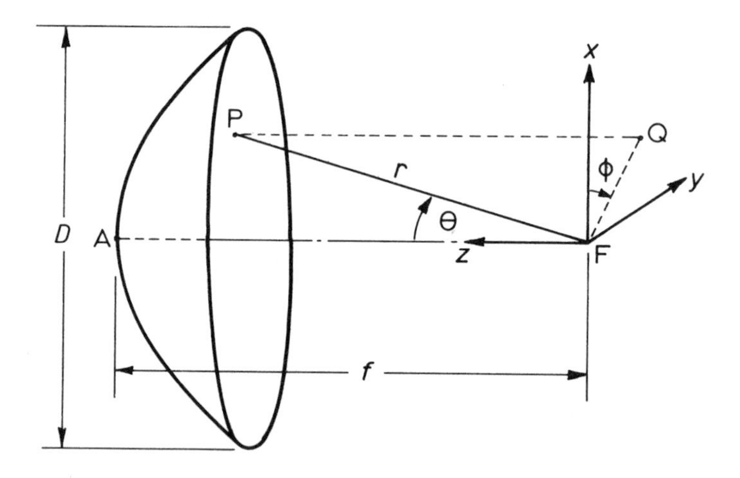

Fig. 7.5 – Paraboloidal reflector illuminated by a small horn at its focus.

The reflector can be assumed to be in the far-field of the feed, since the dimensions of the feed are small. The field incident on the reflector at point P is therefore (equation (3.61)

$$\mathbf{E}_{\text{inc}}(r,\theta,\phi) = j2\pi \frac{\exp(-jkr)}{kr} F_x^{\text{F}}(\alpha,\beta)\,[\,\mathbf{u}_x\gamma - \mathbf{u}_z\alpha\,] \;, \tag{7.17}$$

which is a spherical wave whose pattern is defined by the angular spectrum $F_x^{\text{F}}(\alpha,\beta)$ of the feed, with $\alpha = \sin\theta\cos\phi$, $\beta = \sin\theta\sin\phi$. It was shown in the second example in section 6.3 that such a spherical wavefront, centred on the focus of the paraboloid, is transformed at every point of the surface into a locally planar wavefront perpendicular to the paraboloid axis. In the language of geometrical optics, the bundle of conical rays from F is converted on reflection at P into a cylindrical bundle of rays which travel parallel to the axis AF to intersect the focal plane at Q. The basic simplifying assumption of the projected aperture-field method is that the field at P travels to Q in the manner of geometrical optics (that is, no diffraction takes place) and so the field at Q has the same amplitude and polarization as it has just after reflection at P, but its phase is retarded by an amount corresponding to the distance PQ. The focal plane will be taken to be the aperture plane, from which the field is allowed subsequently to diffract. A better choice would obviously be the plane coinciding with the rim of the reflector, but the choice is made here for mathematical convenience.

The field just after reflection at P is given by equation (6.3) as

$$\mathbf{E}_{\text{refl}} = -\mathbf{E}_{\text{inc}} + 2(\mathbf{n} \cdot \mathbf{E}_{\text{inc}})\mathbf{n} \tag{7.18}$$

where $\mathbf{n}$ is the inward normal to the reflecting surface at P. (The justification for treating the amplitude and phase of the reflected field as though reflection had taken place at an infinite tangent-plane at P (see section 6.3) assumes that the radii of curvature of the surface at P are both large compared to the wavelength). By simple geometry, the normal $\mathbf{n}$ has the components

$$\begin{aligned} n_x &= -\sin(\theta/2)\cos\phi, \\ n_y &= -\sin(\theta/2)\sin\phi, \\ n_z &= -\cos(\theta/2). \end{aligned} \tag{7.19}$$

Then from equations (7.17) to (7.19) and including the extra phase retardation by using equation (7.13), the aperture field over the focal plane is

$$\mathbf{E}_a(Q) = -j2\pi\,\frac{\exp(-j2kf)}{kr}\,F_x^{\,\mathrm{F}}(\alpha,\beta)$$

$$[\mathbf{u}_x(\cos\theta + 2\sin^2(\theta/2)\cos^2\phi) + \mathbf{u}_y\sin^2(\theta/2)\sin 2\phi] \tag{7.20}$$

within a circle of diameter D, and zero outside. The vector part of this aperture-field expression reveals that, whereas the field incident on the reflector was purely x-polarized, the aperture field after reflection contains a y-component. There is no z-component in the aperture field, which is consistent with the assumption that the fields between the reflector and the aperture plane are those of a plane wave travelling parallel to the axis. The electric field lines in the aperture plane are shown in Fig. 7.6, for a projected aperture where the rim of the reflector coincides with the focal plane (that is, $f/D = 0.25$). Apertures with smaller diameters, and hence larger f/D ratios, can be drawn on the same figure to give their projected-aperture field lines, as shown for the case of $f/D = 1$.

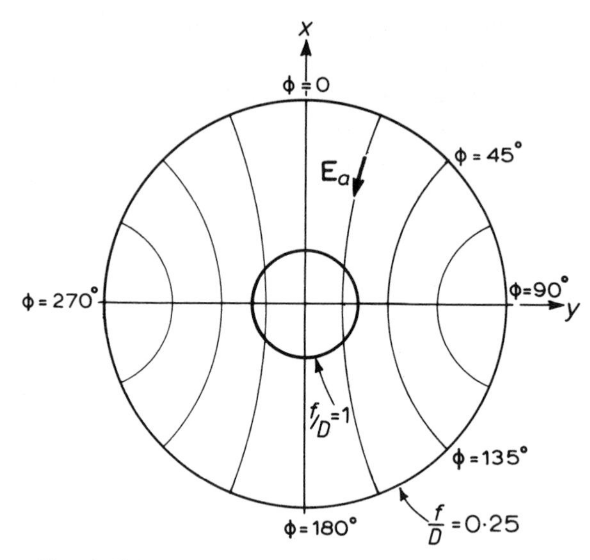

Fig. 7.6 – Electric-field lines on the projected aperture of an axisymmetrically fed paraboloid reflector with an x-polarized horn feed at its focus.

The correspondence between the rectangular coordinates (x,y) of the point Q in the focal plane and the spherical-polar coordinates of the point P on the reflector is given by

$$x = r\alpha = r \sin\theta \cos\phi$$
$$y = r\beta = r \sin\theta \sin\phi \qquad (7.21)$$

with

$$r = \frac{2f}{1 + \cos\theta} = \frac{f}{\cos^2(\theta/2)} \qquad (7.22)$$

Hence the aperture field over the projected aperture, with y replaced by $-y$ in order to describe radiation out into the medium away from the antenna, is in rectangular coordinates

$$\mathbf{E}_a(x,y) = -j2\pi \frac{\exp(-j2kf)}{kr} F_x{}^{\mathrm{F}}\left(\frac{x}{r}, -\frac{y}{r}\right)$$
$$\left[\mathbf{u}_x \left(\frac{2f-r}{r} + \frac{x^2}{2fr}\right) + \mathbf{u}_y \frac{xy}{2fr}\right] \qquad (7.23)$$

with

$$r = f\left(1 + \frac{x^2 + y^2}{4f^2}\right) \qquad (7.24)$$

Alternatively, in circular-polar coordinates (ρ,ϕ), the projected aperture field in this case is

$$\mathbf{E}_a(\rho,\phi) = -j2\pi \frac{\exp(-j2kf)}{kr} F_x{}^{\mathrm{F}}\left(\frac{\rho \cos \phi}{r}, -\frac{\rho \sin \phi}{r}\right)$$

$$\left[\mathbf{u}_x \left(\frac{2f-r}{r} + \frac{\rho^2 \cos^2 \phi}{2fr}\right) + \mathbf{u}_y \frac{\rho^2 \sin 2\phi}{4fr}\right] \tag{7.25}$$

with

$$r = f\left(1 + \frac{\rho^2}{4f^2}\right) \quad . \tag{7.26}$$

Note that the $1/r$ term in the amplitude provides a small natural taper of the aperture field, in addition to that caused by the primary feed pattern.

Before we calculate the diffraction field from this projected aperture field, it is interesting to compare this aperture field with that produced when the paraboloid reflector is illuminated by an elementary dipole. The electric field incident at the dish would then be given by equation (3.114), assuming the current element to be x-directed and of length a, and positioned at the focus (see Fig. 7.5) in place of the horn. Then, following the same procedure as for the x-polarized horn feed (except that there is now a y-polarized component in the illuminating field) the aperture field at Q (cf. equation (7.20)), for illumination by an elementary dipole, is

$$\mathbf{E}_a(Q) = j2\pi \frac{\exp(-j2kf)}{kr} \frac{ZIa}{2\lambda^2}$$

$$[\mathbf{u}_x(1 - 2\sin^2(\theta/2)\cos^2\phi) - \mathbf{u}_y \sin^2(\theta/2)\sin 2\phi] \quad . \tag{7.27}$$

The electric-field lines of this aperture field are sketched in Fig. 7.7.

As in the previous figure, the outer circle is for the situation when the rim of the paraboloid coincides with the focal plane, for which $D = 4f$. The inner circle is drawn for $D = f$.

Returning to the horn-fed paraboloid, whose projected aperture field is given by equation (7.25), we now calculate its far-field radiation pattern. For simplicity we assume that the ratio f/D is large, and that the angular spectrum of the feed is uniform over the reflector. (These are oversimplifications in practice,

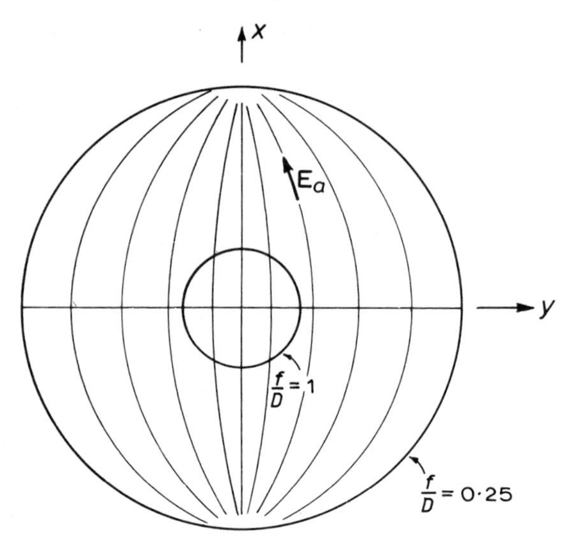

Fig. 7.7 – Electric-field lines on the projected aperture of an axisymmetrically fed paraboloid reflector with an x-directed elemental dipole feed at its focus.

but enable us to obtain some analytical feeling for what is happening.) The projected aperture field is then

$$\mathbf{E}_a(\rho',\phi') = \mathbf{u}_x\, E_{ax}(\rho',\phi') + \mathbf{u}_y\, E_{ay}(\rho',\phi') \tag{7.28}$$

with

$$E_{ax}(\rho',\phi') = E_o \tag{7.29}$$

and

$$E_{ay}(\rho',\phi') = E_o \frac{\rho'^2}{4f^2} \sin 2\phi' \tag{7.30}$$

over the circle of diameter D. Primes are now used on the aperture coordinates (ρ',ϕ') in order to distinguish them from the coordinates of the diffracted field (r,θ,ϕ). The constant

$$E_o = -j2\pi \frac{\exp(-j2kf)}{kf} F_x{}^{\mathrm{F}}(0,0) \quad . \tag{7.31}$$

The angular spectrum of the x-polarized part of this aperture field (see section 3.2.2) is

$$F_x(\alpha,\beta) = E_o \frac{\pi D^2}{4\lambda^2} \frac{2J_1\left(\dfrac{\pi D s}{\lambda}\right)}{\left(\dfrac{\pi D s}{\lambda}\right)} \tag{7.32}$$

where $s = \sin\theta$, and the spectrum is independent of ϕ. The angular spectrum of the y-polarized part is

$$F_y(\alpha,\beta) = \frac{E_o}{4\lambda^2 f^2}\int_0^{D/2} d\rho' \int_0^{2\pi} d\phi'\, \rho'^3 \sin 2\phi'$$

$$\exp\{jk\rho' \sin\theta \cos(\phi-\phi')\} \quad . \tag{7.33}$$

In order to evaluate this double integral the following results are needed. They are couched in fairly general terms for use later on. (A useful collection of results on Bessel functions can be found in Jahnke & Emde (1945)). First, a standard result from the theory of Bessel functions, which is familiar to electrical engineers (see Brown and Glazier (1964), for example, as giving the spectrum of a sinusoidally phase-modulated carrier, is

$$\exp\{jz\cos(\phi-\phi')\} = \sum_{n=-\infty}^{\infty} (j)^n J_n(z) \exp\{jn(\phi-\phi')\} \tag{7.34}$$

where n is an integer, z is real, and $J_n(\,)$ is the Bessel function of the first kind and order n. Multiplying by $\exp(jp\phi')$, where p is also an integer, and then integrating with respect to ϕ' over the range from 0 to 2π,

$$\int_0^{2\pi} \exp\{jz\cos(\phi-\phi')\} \exp(jp\phi')\, d\phi'$$

$$= \sum_{n=-\infty}^{\infty} (j)^n J_n(z) \exp(jn\phi) \int_0^{2\pi} \exp\{j(p-n)\phi'\}\, d\phi' \tag{7.35}$$

The individual terms in this summation will be zero, unless $n = p$, when

$$\int_0^{2\pi} \exp(j0)\, d\phi' = 2\pi$$

Hence

$$\int_0^{2\pi} \exp\{jkz\cos(\phi-\phi')\} \exp(jp\phi')\, d\phi'$$

$$= 2\pi(j)^p J_p(z) \exp(jp\phi) \tag{7.36}$$

The particular result then follows that

$$\int_0^{2\pi} \exp\{jz\cos(\phi-\phi')\}\sin 2\phi'\,d\phi' = -2\pi J_2(z)\sin 2\phi \quad , \tag{7.37}$$

using the relation

$$J_{-n}(z) = (-1)^n J_n(z) \quad . \tag{7.38}$$

The ϕ'-integration of equation (7.33) can now be performed, using equation (7.36). It yields

$$F_y(\alpha,\beta) = -\frac{2\pi E_o}{4\lambda^2 f^2}\sin 2\phi \int_0^{D/2} J_2(k\rho'\sin\theta)\,\rho'^3\,d\rho' \quad . \tag{7.39}$$

We now need the further general result from the theory of Bessel functions, that

$$\int z^{n+1} J_n(z)\,dz = z^{n+1} J_{n+1}(z) \quad . \tag{7.40}$$

Hence equation (7.39) becomes

$$F_y(\alpha,\beta) = -\frac{1}{8}\left(\frac{D}{f}\right)^2 E_o \frac{\pi D^2}{4\lambda^2} \frac{J_3\left(\frac{\pi Ds}{\lambda}\right)}{\left(\frac{\pi Ds}{\lambda}\right)} \sin 2\phi \quad . \tag{7.41}$$

Exercise

Determine the angular spectrum due to the $\cos^2\phi$ term, discarded from the aperture field of equation (7.25). Continue to make the assumption that $r \approx f$.

The x-polarized spectrum of equation (7.32) is clearly at a higher level than the y-polarized spectrum of equation (7.41). This is mainly due to the assumption that the ratio f/D is large. The dependence of the amplitude of the Bessel functions on $u = \pi Ds/\lambda$ is shown in Fig. 7.8. The x-polarized spectrum is circularly symmetic (that is, independent of ϕ), whereas the y-polarized spectrum varies as $\sin 2\phi$, and so has lobes in the $\phi = \pm 45°$ planes. At smaller f/D ratios, when the dependence of r on ρ must be accounted for, and for non-uniform illumination of the reflector by the primary feed, the calculations become more elaborate, and in many cases must be performed numerically. But the above example, despite the unrealistic assumptions, does provide valuable insight into the character of the radiation from an axisymmetrically horn-fed paraboloidal reflector antenna.

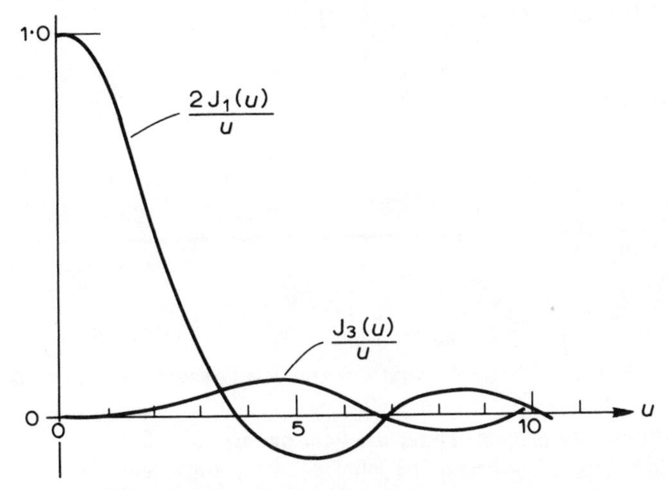

Fig. 7.8 – Dependence of the θ-dependent functions $2J_1(u)/u$ and $J_3(u)/u$ on $u = \pi D \sin\theta/\lambda$.

The natural amplitude taper of the aperture field (see equations (7.25) and (7.26)) due to the dependence of r on ρ, which was ignored in the above calculations, has the generally beneficial effect of reducing the level of the sidelobes. There is a corresponding slight reduction in gain (see section 4.1.2) and a slight broadening in the width of the main beam (see Exercise p. 64). This natural taper is often enhanced by deliberately tapering the feed illumination from the centre to the rim of the reflector.

Exercise
Show that, for modest f/D ratios greater than unity, a reasonable approximation to the natural amplitude taper in the aperture field of equation (7.25) is the parabolic variation $1 - \rho^2/(2f^2)$.

Exercise
Calculate the angular spectrum for the aperture field

$$E_{ax}(\rho',\phi') = E_0\left[1 - \left(\frac{\rho'}{a}\right)^2\right]\mathrm{circ}_D(\rho') \tag{7.42}$$

where $a \geqslant D/2$, and the circle function is defined as

$$\mathrm{circ}_D(\rho') = \begin{cases} 1 \text{ for } \rho' \leqslant D/2 \\ 0 \text{ for } \rho' > D/2 \end{cases} \quad . \tag{7.43}$$

It may be helpful to note that, as a consequence of the recurrence relationship for Bessel functions of different orders

$$J_{n-1}(z) + J_{n+1}(z) = \frac{2n}{z} J_n(z) \tag{7.44}$$

we have, for $n = 1$,

$$J_0(z) + J_2(z) = \frac{2}{z} J_1(z) \tag{7.45}$$

and hence that

$$z^3 J_0(z) = 2z^2 J_1(z) - z^3 J_2(z) \quad . \tag{7.46}$$

Hence estimate the effect on the first sidelobe level of various degrees of taper.

7.2.3 Validity of the projected aperture-field method

The question must be asked as to whether the geometrical-optics assumption used in constructing the aperture field introduces unacceptable errors in estimating the radiated field. If the reflector is in the far field of the feed, the assumption is justified right up to the reflector surface. This requires only that the dimensions of the feed be small enough, according to the Rayleigh far-field criterion. But the next step in the argument, which stipulates that the field travels from the paraboloidal reflector surface to the aperture plane in parallel-sided ray tubes, is the one about which there might be some doubt.

The far field calculated from the focal-plane aperture field of equation (7.25), which is of the general form

$$\mathbf{E}_a(\rho',\phi') = \mathbf{u}_x E_{ax}(\rho',\phi') + \mathbf{u}_y E_{ay}(\rho',\phi') \tag{7.47}$$

over the circular aperture of diameter D, and zero outside, is given by equations (3.80) and (3.81), or by (3.82) and (3.83), with equations (3.76) and (3.77) substituted. Restricting attention for the moment to the x-polarized component of the aperture field, and transforming the aperture coordinates from rectangular (x,y) to circular-polar (ρ',ϕ'), the far field is

$$\mathbf{E}(r,\theta,\phi) = j \frac{2\pi}{\lambda^2} \frac{\exp(-jkr)}{kr} [\mathbf{u}_\theta \cos\phi - \mathbf{u}_\phi \cos\theta \sin\phi]$$

$$\int_0^{D/2} \mathrm{d}\rho' \int_0^{2\pi} \mathrm{d}\phi' \, \rho' E_{ax}(\rho',\phi') \exp\{jkr \sin\theta \cos(\phi-\phi')\} \tag{7.48}$$

in the geometry of Fig. 7.9. The y-polarized component of the aperture field can be treated similarly. It should be stressed that the following argument applies to any focal-plane aperture field, provided the paraboloidal reflector is in the far field of the primary feed at its focus.

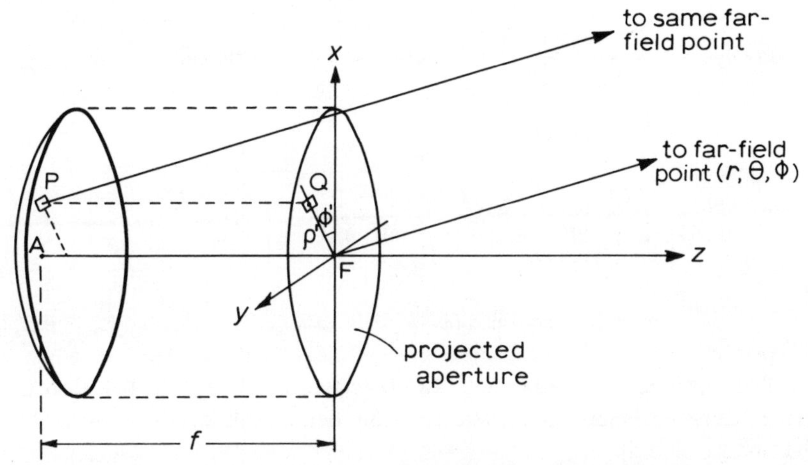

Fig. 7.9 – Geometry of the projected aperture field in the focal plane.

Now suppose that the element $\rho' \mathrm{d}\rho' \mathrm{d}\phi'$ in the focal-plane projected aperture is, as it were, taken back on to the paraboloidal surface but without changing its orientation. That is, it is an equivalent projected aperture *at* the surface (cf. Fig. 6.5) whose aperture field is, by construction, precisely that at Q in amplitude and polarization, but advanced in phase by an amount corresponding to the distance PQ. Thus the equivalent projected aperture field at P is

$$\mathbf{E}_a(\mathrm{P}) = \mathbf{E}_a(\rho',\phi') \exp\left\{jk\left(f - \frac{\rho'^2}{4f}\right)\right\} \tag{7.49}$$

The far field for this elemental field will now be calculated, and an integration performed over all such projected aperture-field elements to obtain the total far field.

Inserting this elemental-field expression (7.49) into the general far-field form of Huygens' principle (equation 3.100 and Fig. 3.11)

$$\begin{aligned} \mathrm{d}\mathbf{E}(r,\theta,\phi)_{r\to\infty} = j\, & \frac{2\pi}{\lambda^2} \frac{\exp(-jkr)}{kr} \rho' \mathrm{d}\rho' \mathrm{d}\phi' \\ & \exp\left\{jk(1-\gamma)\left(f - \frac{\rho'^2}{4f}\right)\right\} \\ & \exp\{jk\rho' \sin\theta \cos(\phi - \phi')\} \\ & [E_{ax}(\rho',\phi')(\mathbf{u}_x\gamma - \mathbf{u}_z\,\alpha) + E_{ay}(\rho',\phi')(\mathbf{u}_y\gamma - \mathbf{u}_z\beta)] \end{aligned} \tag{7.50}$$

If the y-component of the aperture field is discarded, just for the purposes of comparison with equation (7.48), and equation (7.50) is integrated over the entire projected aperture, the total far field is given by the more accurate expression (converting to spherical-polar components)

$$\mathbf{E}(r,\theta,\phi) = j\,\frac{2\pi}{\lambda^2}\;\frac{\exp(-jkr)}{kr}\,[\mathbf{u}_\theta \cos\phi - \mathbf{u}_\phi \cos\theta \sin\phi]$$

$$\exp\{jkf(1-\gamma)\}$$

$$\int_0^{D/2} \mathrm{d}\rho' \int_0^{2\pi} \mathrm{d}\phi'\, \rho'\, E_{ax}(\rho',\phi') \exp\left\{-jk(1-\gamma)\frac{\rho'^2}{4f}\right\}$$

$$\exp\{jkr\sin\theta\cos(\phi-\phi')\} \quad . \tag{7.51}$$

This formula for the far field differs in two respects from that derived by way of the focal-plane projected aperture field, both of them phase terms. The first phase term,

$$\exp\{jkf(1-\gamma)\} \tag{7.52}$$

which is outside the integral sign, and therefore easy to make allowance for, arises from the change in phase reference from A to F. The angle-independent factor $\exp(jkf)$ simply reflects the phase differences for ray path AF. The angle-dependent factor $\exp(-jkf\gamma)$ arises from a change in the phase-reference origin, with respect to which the angular plane-wave spectrum is defined, from A to F. (This is explained in section 5.1 and Fig. 5.2). The second phase term is a parabolic phase term inside the integral sign:

$$\exp\left\{-jk(1-\gamma)\frac{\rho'^2}{4f}\right\} \quad . \tag{7.53}$$

For accurate work it must of course be incorporated in the integration. But it is interesting to know in what circumstances it can be ignored. Using the same arguments that were used in deriving the Rayleigh far-field criterion (section 5.3.1), this term can be set equal to unity provided the maximum phase excursion is $\pi/8$ or less. Thus

$$1-\gamma \leqslant \frac{\lambda f}{D^2} \tag{7.54}$$

which means, for small θ, that

$$\theta \leqslant \frac{\sqrt{(2\lambda f)}}{D} \tag{7.55}$$

for the parabolic phase term to have a negligible effect. Since the 3-dB beamwidth of a uniform circular aperture is approximately λ/D (see Fig. 3.7), equation (7.55) can be interpreted as meaning that the focal-plane projected aperture-field technique of pattern analysis is approximately valid out to angles which are $\sqrt{(2f/\lambda)}$ times the 3-dB beamwidth from the (boresight) z-axis. So the aperture-field technique is perfectly acceptable for calculating the magnitude of the radiation pattern out to angles of several times the 3-dB beamwidth.

7.2.4 Far field calculated by a surface integration method

A more precise method of analysing the far field of the radiation from an axisymmetrically fed paraboloidal reflector antenna will now be given, based on the method of 'reflecting Huygens elements' introduced in section 6.4.1. Each element of the reflecting surface is treated as a radiating Huygens source, the source field being derived from the incident field by way of the tangential electric-field boundary condition. Integration over the surface of the reflector yields the total radiated field. The method is basically more precise than either the projected aperture-field method described in section 7.2.2, or its modification in section 7.2.3 which used an essentially physical argument to test the validity of the aperture-field method. The method of surface integration of reflecting Huygens elements is equivalent to the conventional technique of integrating over the surface currents induced by the illuminating field and assuming that the field behaves locally at the reflector as if reflected by the tangent plane (see Silver (1949) and Harrington (1959 and 1961)).

The geometry appropriate to the application of the reflecting Huygens elements method is that of Fig. 7.10.

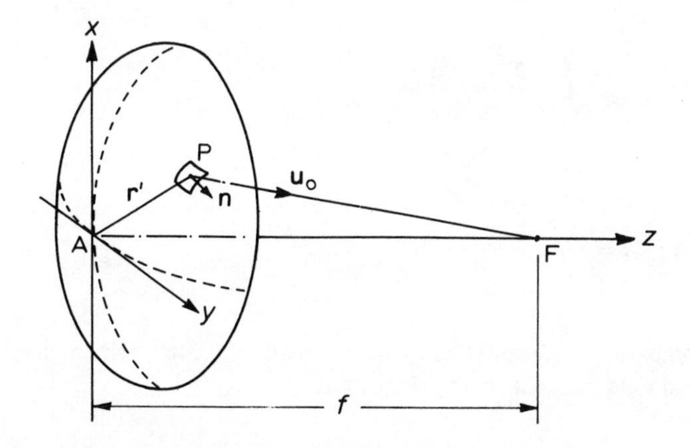

Fig. 7.10 – A surface element at P, on the surface of a paraboloidal reflector, viewed from the rear side.

To use the results of section 6.4.1, the surface element at P is the projection of an element $dx\,dy$ in the apical plane (that is, the x-y plane through the apex A, which is taken as corresponding to the reference point O of Fig. 6.9). The vector position of P is

$$\mathbf{r}' = \mathbf{u}_x x + \mathbf{u}_y y + \mathbf{u}_z \frac{x^2 + y^2}{4f} \tag{7.56}$$

and the direction ratios for the unit normal vector $\mathbf{n}$ at P are (from equations 6.42 to 6.44)

$$\frac{n_x}{nz} = -\frac{x}{2f} \quad \text{and} \quad \frac{n_y}{n_z} = -\frac{y}{2f} \tag{7.57}$$

The field incident at P will be assumed to come from an x-polarized primary radiator whose dimensions are sufficiently small for all parts of the surface to be considered in its far field. Equation (7.17) gives the details of this spherical wave. But at P the wave can be taken to be locally a plane wave arriving from the direction $\mathbf{u}_o$, which is the direction of the focus viewed from P. Thus

$$\mathbf{u}_o = \mathbf{u}_x\,\alpha_o + \mathbf{u}_y\,\beta_o + \mathbf{u}_z\,\gamma_o \quad , \tag{7.58}$$

where (with $r =$ FP)

$$\begin{aligned} \alpha_o &= -x/r \\ \beta_o &= -y/r \\ \gamma_o &= \left(f - \frac{x^2 + y^2}{4f}\right)\Big/ r \end{aligned} \tag{7.59}$$

and

$$r = f + \frac{x^2 + y^2}{4f} \tag{7.60}$$

Comparing equation (7.17) with equation (6.46), which denotes the x-polarized field incident on the element as

$$\mathbf{e}_{ox} = e_{ox}\left(\mathbf{u}_x - \mathbf{u}_z \frac{\alpha_o}{\gamma_o}\right) \tag{7.61}$$

we have that

$$e_{ox} = j2\pi \frac{\exp(-jkr)}{kr} \exp(-jk\mathbf{u}_o . \mathbf{r}')$$

$$\gamma_o F_x{}^F \left(\frac{x}{r}, -\frac{y}{r}\right) \tag{7.62}$$

referred to A as phase reference.

Equations (6.55) to (6.57) give the equivalent aperture field, corresponding to the far field calculated by integrating over the reflecting Huygens elements, as

$$\mathbf{E}_a(x,y) = -e_{ox} \exp\{jk\,[\alpha_o\, x + \beta_o\, y + (\gamma_o + \gamma)\zeta(x,y)]\}$$

$$\left[\mathbf{u}_x\left(1 + \frac{n_x}{n_y}\frac{\alpha_o}{\gamma_o} + \frac{n_y}{n_z}\frac{\beta}{\gamma}\right) + \mathbf{u}_y\frac{n_y}{n_z}\left(\frac{\alpha_o}{\gamma_o} - \frac{\alpha}{\gamma}\right)\right] \tag{7.63}$$

in which $\zeta(x,y)$ is the profile of the reflecting surface.

Equation (7.63) becomes, when all the preceding equations are substituted,

$$\mathbf{E}_a(x.y) = -j2\pi \frac{\exp(-jkr)}{kr} F_x{}^F\left(\frac{x}{r}, -\frac{y}{r}\right) \exp\left\{jk\gamma\frac{x^2+y^2}{4f}\right\}$$

$$\left[\mathbf{u}_x\left\{\frac{f - \dfrac{x^2+y^2}{4f}}{r} + \frac{x^2}{2fr} - \frac{y\left(f - \dfrac{x^2+y^2}{4f}\right)}{2fr}\frac{\beta}{\gamma}\right\}\right.$$

$$\left. + \mathbf{u}_y\left\{\frac{xy}{2fr} - \frac{y\left(f - \dfrac{x^2+y^2}{4f}\right)}{2fr}\frac{\alpha}{\gamma}\right\}\right] \tag{7.64}$$

Close examination of this aperture-field formula reveals that two new terms have appeared, both principally dependent on y. Their dependence on α and β suggests that they will be unimportant in the direction of the main beam, but they may well become significant in directions well away from boresight. If these terms are ignored, and if the f/D ratio is somewhat larger than unity, equation (7.64) is essentially the modified projected aperture field derived previously using physical arguments.

Exercise

Make an exact transformation of the aperture field of equation (7.64) from the apical to the focal plane. Hence justify the statement made in the final sentence of the last paragraph.

7.2.5 Fields in the focal plane of a paraboloid reflector axially illuminated by a plane wave

If the concave paraboloidal surface of Fig. 7.10 is illuminated by a plane wave incident along its axis we have, in effect, a uniform circular aperture field focused on the point F. Consideration of a focused one-dimensional aperture field, treated in section 5.3.2, would lead us to expect the field distribution in the focal plane to be of the form $J_1(u)/u$ where u is proportional to the radial coordinate. That this is indeed so will now be demonstrated using the approximate equivalent aperture field of section 6.3 together with Fresnel's diffraction formula for three-dimensional fields (from section 5.5.1). Later on in this section the more precise equivalent aperture field of section 6.4 will be used to give a fuller picture of the focal-plane fields.

For a shallow paraboloidal reflector, whose f/D ratio is large, with an x-polarized plane wave incident along its axis, an adequate approximation for its equivalent aperture field is

$$E_{ax}(x,y) = -E_o \exp\left\{jk\frac{x^2+y^2}{2f}\right\} \tag{7.65}$$

over the circular aperture of diameter D and zero outside. Cartesian coordinates will be used initially, and the finite domain of the aperture field in the x-y plane indicated by the symbol $\circ$. If the diameter D of the reflector is large in comparison with the wavelength the corresponding angular spectrum of the reflected radiation will be fairly narrowly confined around the direction of the z-axis, and the Fresnel diffraction formula of equation (5.84) can be used. This gives the field in the focal plane, on substituting equation (7.65), as

$$E_x(x,y,f) = -\frac{jE_o}{\lambda f}\exp\left\{-jk\left(f+\frac{x^2+y^2}{2f}\right)\right\}$$

$$\iint\limits_{\circ} \exp\left\{jk\left(\frac{x}{f}x'+\frac{y}{f}y'\right)\right\}\,dx'dy' \quad . \tag{7.66}$$

In terms of polar coordinates (ρ,ϕ) in the focal plane, and (ρ',ϕ') in the aperture plane, this becomes

$$E_x(\rho,\phi,f) = -\frac{jE_o}{\lambda f} \exp\left\{-jk\left(f + \frac{\rho^2}{2f}\right)\right\}$$

$$\int_0^{D/2}\int_0^{2\pi} \exp\left\{jk\frac{\rho}{f}\rho' \cos(\phi-\phi')\right\} \rho' d\rho' d\phi' \quad . \qquad (7.67)$$

The integrations are straightforward, using equations (7.36) and (7.40), so that

$$E_x(\rho,\phi,f) = -j\frac{E_o\pi(D/2)^2}{\lambda f} \exp\left\{-jk\left(f + \frac{\rho^2}{2f}\right)\right\} \frac{2J_1\left(\dfrac{\pi\rho D}{\lambda f}\right)}{\left(\dfrac{\pi\rho D}{\lambda f}\right)} \qquad (7.68)$$

which has the expected dependence on ρ and is independent of ϕ. Note that this focal-plane field is of precisely the same form as the far field, given by equation (7.32), when the same shallow concave paraboloidal reflector is illuminated uniformly from its focus.

The more precise equivalent aperture field, given by equations (6.55) to (6.57), for the concave paraboloid illuminated by an x-polarized plane wave incident along its axis is

$$\mathbf{E}_a(x,y) = -E_o \exp\left\{jk(1+\gamma)\frac{x^2+y^2}{4f}\right\}$$

$$\left[\mathbf{u}_x\left(1 - \frac{y}{2f}\frac{\beta}{\gamma}\right) + \mathbf{u}_y\frac{y}{2f}\frac{\alpha}{\gamma}\right] \quad . \qquad (7.69)$$

It can be seen from this that the aperture field of equation (7.65) is a good approximation when the ratios of f/D and D/λ are both large compared to unity, and $\gamma \approx 1$ in the Fresnel approximation. But it is interesting to calculate the y-component of the electric field in the focal plane. This will be done by extending slightly the Fresnel diffraction analysis of section 5.5.1.

From the fundamental formula (3.78) describing the fields in $z \geqslant 0$ (exactly) in terms of their angular spectra, the y-component of the electric field in the focal plane $z = f$ is

$$E_y(x,y,f) = \int_{-\infty}^{\infty}\int_{-\infty}^{\infty} F_y(\alpha,\beta)$$

$$\exp\{-jk(\alpha x + \beta y + \gamma z)\}\, d\alpha d\beta \qquad (7.70)$$

with

$$F_y(\alpha,\beta) = \frac{1}{\lambda^2} \int_{-\infty}^{\infty} \int_{-\infty}^{\infty} E_{ay}(x,y) \exp \{jk(\alpha x + \beta y)\} \, dx dy \tag{7.71}$$

Assuming that this spectrum is narrowly confined around the direction of the z-axis, $\gamma \approx 1 - \frac{1}{2}(\alpha^2 + \beta^2)$, and equation (7.70) takes the approximate form

$$E_y(x,y,f) = \exp(-jkf) \int_{-\infty}^{\infty} \int_{-\infty}^{\infty} F_y(\alpha,\beta) \exp\left\{\frac{jkf}{2}(\alpha^2 + \beta^2)\right\}$$

$$\exp\{-jk(\alpha x + \beta y)\} \, d\alpha d\beta \tag{7.72}$$

To the same degree of approximation the y-component of the aperture field of equation (7.69) is

$$E_{ay}(x,y) = -E_o \frac{y}{2f} \alpha \exp\left\{jk \frac{x^2 + y^2}{2f}\right\} \tag{7.73}$$

$$= \frac{\alpha y}{2f} E_{ax}(x,y) \quad , \tag{7.74}$$

where $E_{ax}(x,y)$ is the approximate form for the x-component of the aperture field of equation (7.69).

Now equation (7.72) is the inverse two-dimensional Fourier transform of the product of two functions of α and β. Using the symbol $\mathscr{F}[\ ]$ to denote a Fourier transform such as that of equation (7.71), which would then be written symbolically as

$$F_y(\alpha,\beta) = \mathscr{F}[E_{ay}(x,y)] \tag{7.75}$$

and its inverse as

$$E_{ay}(x,y) = \int_{-\infty}^{\infty} \int_{-\infty}^{\infty} F_y(\alpha,\beta) \exp\{-jk(\alpha x + \beta y)\} \, d\alpha d\beta$$

$$= \mathscr{F}^{-1}[F_y(\alpha,\beta)] \quad , \tag{7.76}$$

we can write equation (7.72) as

$$E_y(x,y,f) = \exp(-jkf) \mathscr{F}^{-1}\left[F_y(\alpha,\beta) \exp\left\{\frac{jkf}{2}(\alpha^2 + \beta^2)\right\}\right] \tag{7.77}$$

From equations (7.74) and (7.75)

$$F_y(\alpha,\beta) = \frac{\alpha}{2f}\mathscr{F}[y\,E_{ax}(x,y)] \tag{7.78}$$

and so equation (7.77) can be rewritten as

$$E_y(x,y,f) = \frac{1}{2f}\exp(-jkf)\mathscr{F}^{-1}\left[\mathscr{F}[y\,E_{ax}(x,y)]\;\alpha\exp\left\{\frac{jkf}{2}(\alpha^2+\beta^2)\right\}\right] \tag{7.79}$$

The standard integral of equation (6.64) can be used to show that

$$\mathscr{F}^{-1}\left[\exp\left\{\frac{jkf}{2}(\alpha^2+\beta^2)\right\}\right] = \frac{j\lambda}{f}\exp\left\{-\frac{jk}{2f}(x^2+y^2)\right\} \tag{7.80}$$

It also follows (for example, by differentiating equation (7.76)) partially with respect to x) that

$$\mathscr{F}^{-1}\left[\alpha\exp\left\{\frac{jkf}{2}(\alpha^2+\beta^2)\right\}\right] = \frac{j}{k}\frac{\partial}{\partial x}\mathscr{F}^{-1}\left[\exp\left\{\frac{jkf}{2}(\alpha^2+\beta^2)\right\}\right] \tag{7.81}$$

and hence that

$$\alpha\exp\left\{\frac{jkf}{2}(\alpha^2+\beta^2)\right\} = j\frac{\lambda}{f^2}\mathscr{F}\left[x\exp\left\{-\frac{jk}{2f}(x^2+y^2)\right\}\right] \tag{7.82}$$

The focal-plane field of equation (7.79) becomes

$$E_y(x,y,f) = j\frac{\lambda}{2f^3}\exp(-jkf)\mathscr{F}^{-1}\left[\mathscr{F}[y\,E_{ax}(x,y)]\;\mathscr{F}\left[x\exp\left\{-\frac{jk}{2f}(x^2+y^2)\right\}\right]\right] \tag{7.83}$$

which is now in a form to which the two-dimensional convolution theorem applies. Hence, using equation (5.83),

$$E_y(x,y,f) = j\,\frac{\lambda}{2f^3}\,\exp(-jkf)\,\frac{1}{\lambda^2}\int_{-\infty}^{\infty}\int_{-\infty}^{\infty} y'\,E_{ax}(x',y')$$

$$(x-x')\exp\left\{-\frac{jk}{2f}\,[(x-x')^2+(y-y')^2]\right\}\,\mathrm{d}x'\mathrm{d}y' \quad . \qquad (7.84)$$

Inserting the aperture field of equation (7.65), and expanding the argument of the exponential in the integrand,

$$E_y(x,y,f) = -\frac{jE_o}{2\lambda f^3}\exp\left\{-jk\left(f+\frac{x^2+y^2}{2f}\right)\right\}$$

$$\iint_{O} y'\,(x-x')\exp\left\{j\frac{k}{f}\,(xx'+yy')\right\}\mathrm{d}x'\mathrm{d}y' \qquad (7.85)$$

and in polar coordinates

$$E_y(\rho,\phi,f) = -\frac{jE_o}{2\lambda f^3}\exp\left\{-jk\left(f+\frac{\rho^2}{2f}\right)\right\}$$

$$\int_0^{D/2}\int_0^{2\pi} \rho'\sin\phi'\,(\rho\cos\phi-\rho'\cos\phi')$$

$$\exp\left\{jk\,\frac{\rho}{f}\,\rho'\cos(\phi-\phi')\right\}\rho'\mathrm{d}\rho'\mathrm{d}\phi' \quad . \qquad (7.86)$$

The integrations can be performed, using equations (7.37) and (7.40), to yield the y-polarized component of the electric field in the focal plane as

$$E_y(\rho,\phi,f) = -jE_o\,\frac{D}{2f}\,\frac{\pi D^2}{4\lambda f}\exp\left\{-jk\left(f+\frac{\rho^2}{2f}\right)\right\}$$

$$\left[\frac{\rho}{f}\,\frac{J_2\left(\dfrac{\pi\rho D}{\lambda f}\right)}{\left(\dfrac{\pi\rho D}{\lambda f}\right)}\cos^2\phi - \frac{D}{4f}\,\frac{J_3\left(\dfrac{\pi\rho D}{\lambda f}\right)}{\left(\dfrac{\pi\rho D}{\lambda f}\right)}\sin 2\phi\right] \qquad (7.87)$$

Exercise

Calculate the focal-plane field due to the second term (neglected until now) in the x-polarized part of the aperture field of equation (7.69).

Exercise
Use the foregoing method to derive the longitudinal component of the electric field in the Fresnel region for an arbitrary x-polarized aperture-field distribution. The result should be compatible with equation (5.84) which gives the lateral component of the electric field in the Fresnel region.

7.2.6 Aperture blocking

We have seen that by suitable feed design the aperture field of the axisymmetrically fed paraboloid reflector can be tapered towards the edge to give a low level of radiation in the sidelobe region. But we have so far ignored the effect of the presence of the feed on the radiation. This effect will now be shown to be, in the main, that the level of the first and subsequent odd sidelobes is increased, and that the gain is decreased.

The feed is likely to have lateral dimensions of a wavelength or two. Suppose that its effect is to cast a shadow on the projected aperture field in the form of a circular disc of diameter d, as shown in Fig. 7.11. The actual shape of the disc will be seen to be fairly unimportant, and so a circular shape is chosen for

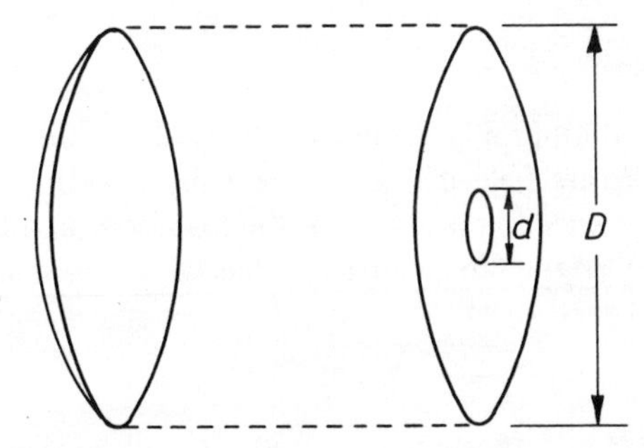

Fig. 7.11 – The projected aperture field blocked by a circular disc representing the feed.

convenience. It has the added advantages of making the analysis applicable in the case of a Cassegrain antenna, when the aperture field is blocked by the secondary reflector. Suppose now that the unblocked aperture field is uniform in ϕ but is dependent on ρ, probably in the form of an amplitude taper, specifically $E_{ax}(\rho)$. Then the analysis of section 3.2.2 gives the angular spectrum of the blocked aperture field as

$$F_x(s) = \frac{2\pi}{\lambda^2}\int_{d/2}^{D/2} E_{ax}(\rho) J_0(k\rho s)\, \rho \mathrm{d}\rho$$

$$= \frac{2\pi}{\lambda^2}\int_{0}^{D/2} E_{ax}(\rho) J_0(k\rho s)\, \rho \mathrm{d}\rho \quad -$$

$$-\frac{2\pi}{\lambda^2}\int_0^{d/2} E_{ax}(\rho)J_o(k\rho s)\,\rho d\rho \tag{7.88}$$

$$= F_{xo}(s) - \frac{2\pi}{\lambda^2}\int_0^{d/2} E_{ax}(\rho)J_o\,(k\rho s)\,\rho d\rho \tag{7.89}$$

where $F_{xo}(s)$ is the unblocked angular spectrum and $s = \sin\theta$, where θ is the angle to the antenna axis.

It is reasonable to assume that the unblocked aperture field is uniform over the area of the disc of diameter d, in which case

$$F_x(s) = F_{xo}(s) - \frac{2\pi}{\lambda^2}\,E_{ax}(0)\int_0^{d/2} J_o(k\rho s)\,\rho d\rho$$

$$= F_{xo}(s) - \frac{\pi d^2}{4\lambda^2}\,E_{ax}(0)\,\frac{2J_1\left(\frac{\pi ds}{\lambda}\right)}{\left(\frac{\pi ds}{\lambda}\right)} \tag{7.90}$$

by making use of equation (3.39). A further practical point is that the diameter D of the unblocked aperture field is likely to be many wavelengths, whereas d will be only one or two. Hence in the region of the main lobe and first few sidelobes of the unblocked pattern the blocked pattern has the approximate form

$$F_x(s) = F_{xo}(s) - \frac{\pi d^2}{4\lambda^2}\,E_{ax}(0)\ . \tag{7.91}$$

The resulting pattern can be deduced from Fig. 7.12. The modification to the

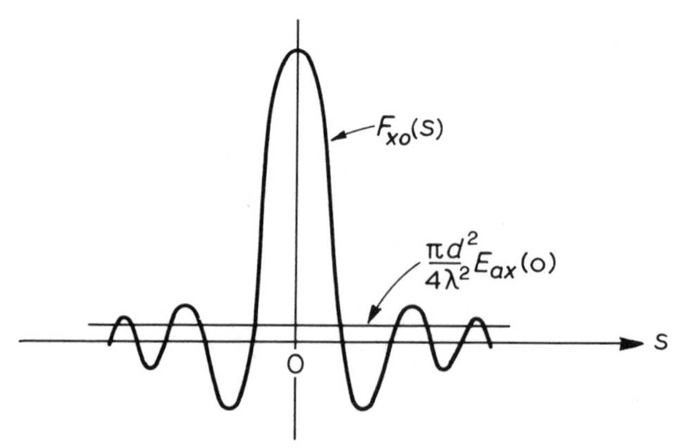

Fig. 7.12 – The blocked angular spectrum is $F_x(s) = F_{xo}(s) - \pi d^2/(4\lambda^2)\,E_{ax}(0)$.

unblocked pattern, prescribed by equation (7.91), is to subtract a small constant value.

The blocked pattern is therefore the unblocked pattern measured to this constant value as a baseline. It is then immediately obvious that the two major effects of aperture blocking are to reduce the gain, since

$$F_x(0) = F_{xo}(0) - \frac{\pi d^2}{4\lambda^2} E_{ax}(0) \quad , \tag{7.92}$$

and to increase the amplitude of the first, third, and subsequent odd sidelobes.

The shadowing effect of supporting struts and other obstructions can be allowed for in a similar manner.

Exercise

Compare the gain and first sidelobe level for a uniform circular aperture field of diameter $D \gg \lambda$ when it is blocked by a circular feed whose diameter d is of the order of the wavelength.

7.2.7 Random reflector-profile effects

If a paraboloid reflector surface is perturbed from its desired form in a random manner, due to manufacturing errors or perhaps, in the case of very large reflectors, to strains in response to local stresses, the wavefront transforming action which normally takes place will be degraded. The main effect of such perturbations of the surface profile will be to introduce local, random, variations of phase into the aperture field. These in turn reduce the design gain, increase the sidelobe levels, and fill in the nulls. The following, mainly physical, analysis indicates how this occurs, and gives an estimate of the reduction in gain.

Fig. 7.13 depicts a small section of the reflector surface whose profile is distorted in a random manner. From a geometrical optics viewpoint the section is illuminated overall by a cone of rays coming from the focus of the paraboloid,

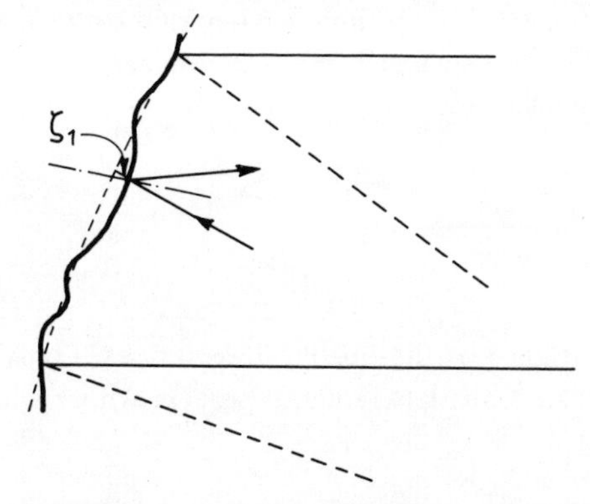

Fig. 7.13 – A section of an irregular reflector.

and for a perfect surface this cone would become a parallel-ray tube. But when the profile is irregular rays are reflected locally in random directions, and the ray picture ceases to be helpful. Indeed, since the lateral scale size of the irregularities could be as small as a wavelength, geometrical optics is not necessarily applicable. (Lateral scale sizes considerably smaller than the wavelength constitute an effective surface reactance which behaves mainly as a perfect reflector.)

However, the method of reflecting Huygens elements (section 6.4.1) gives the following alternative approach. In applying the more precise equivalent aperture field of equations (6.56) and (6.57) to the distorted reflector of Fig. 7.13 we are concerned with the effect of departures from the ideal paraboloidal shape. So if the local slope of the irregular surface is not too different from that of the ideal surface it is evident from equations (6.56) and (6.57) that the principal effect of the irregularities is to introduce a random spatial phase modulation of the ideal aperture field. Thus the actual aperture field can be written to a good approximation as

$$\mathbf{E}_a(x,y) = \mathbf{E}_{ao}(x,y) \exp\{j\Phi(x,y)\} \tag{7.93}$$

where $\mathbf{E}_{ao}(x,y)$ is the equivalent aperture field for the ideal surface profile (which need not be paraboloidal, but whatever the designer had in mind.) The random phase

$$\Phi(x,y) = k(\gamma_o + \gamma)\zeta_1(x,y) \tag{7.94}$$

is linearly related to the departure $\zeta_1\ (x,y)$ of the surface profile from its ideal. The local (cosine of the) angle of incidence γ_o, and the statistics of $\zeta_1(x,y)$ both depend on (x,y), and γ depends on the angle of observation.

We shall now suppose that $\gamma_o + \gamma \approx 2$, which is relevant to large-aperture paraboloidal reflector antennas with large f/D ratios, and that the irregularities are statistically stationary (that is, independent of position). A further simplifying assumption is to take $\zeta_1(x,y)$ as being a zero-mean Gaussian random process, whose probability density function

$$p_{\zeta_1}(z) = \frac{1}{\sqrt{2\pi}\,\sigma_{\zeta_1}} \exp\left\{-\frac{z^2}{2\sigma_{\zeta_1}{}^2}\right\} \tag{7.95}$$

where $\sigma^2_{\zeta_1}$ is the variance of the profile irregularities. Consequently, $\Phi(x,y)$ is a stationary, zero-mean Gaussian random process whose root variance (that is, standard deviation)

$$\sigma_\Phi = 2k\sigma_{\zeta_1} \tag{7.96}$$

The angular spectrum corresponding to the x-component of the aperture field of equation (7.93) is

$$F_x(\alpha,\beta) = \frac{1}{\lambda^2}\int_{-\infty}^{\infty}\int_{-\infty}^{\infty} E_{axo}(x,y)\exp\{j\Phi(x,y)\}$$

$$\exp\{jk(\alpha x + \beta y)\}\,\mathrm{d}x\mathrm{d}y \quad . \qquad (7.97)$$

The expected value (that is, theoretical average or mean) is defined as

$$\langle F_x(\alpha,\beta)\rangle = \int_{-\infty}^{\infty} F_x(\alpha,\beta)\,p_\Phi(\phi)\,\mathrm{d}\phi \quad , \qquad (7.98)$$

where $p_\Phi(\phi)$ is the probability density function of the phase process, and so

$$\langle F_x(\alpha,\beta)\rangle = \frac{1}{\lambda^2}\int_{-\infty}^{\infty}\int_{-\infty}^{\infty}\mathrm{d}x\mathrm{d}y\int_{-\infty}^{\infty}\mathrm{d}\phi\frac{1}{\sqrt{2\pi}\,\sigma_\Phi}\exp\left\{j\phi - \frac{\phi^2}{2\sigma_\Phi^2}\right\}$$

$$E_{axo}(x,y)\,\{jk(\alpha x + \beta y)\} \qquad (7.99)$$

Using the standard integral of equation (2.96), the mean angular spectrum is

$$\langle F_x(\alpha,\beta)\rangle = \exp\{-\tfrac{1}{2}\sigma_\Phi^2\}\,\frac{1}{\lambda^2}$$

$$\int_{-\infty}^{\infty}\int_{-\infty}^{\infty} E_{axo}(x,y)\exp\{jk(\alpha x + \beta y)\}$$

$$= \exp\{-\tfrac{1}{2}\sigma_\Phi^2\}\,F_{xo}(\alpha,\beta) \qquad (7.100)$$

where $F_{xo}(\alpha,\beta)$ is the angular spectrum in the absence of irregularities in the surface profile.

Since $\langle F_x(\alpha,\beta)\rangle$ has the same phase as $F_{xo}(\alpha,\beta)$, and in general only differs from it by a constant value dependent on the probability density of the surface irregularities, the mean angular spectrum describes the coherent part of the radiation. The situation is not unlike that of aperture blocking, as Fig. 7.14 shows, for an aperture field that is ideally circularly symmetric. The coherent power is reduced from the ideal by a constant factor $\exp\{-\sigma_\Phi^2\}$ in every direction.

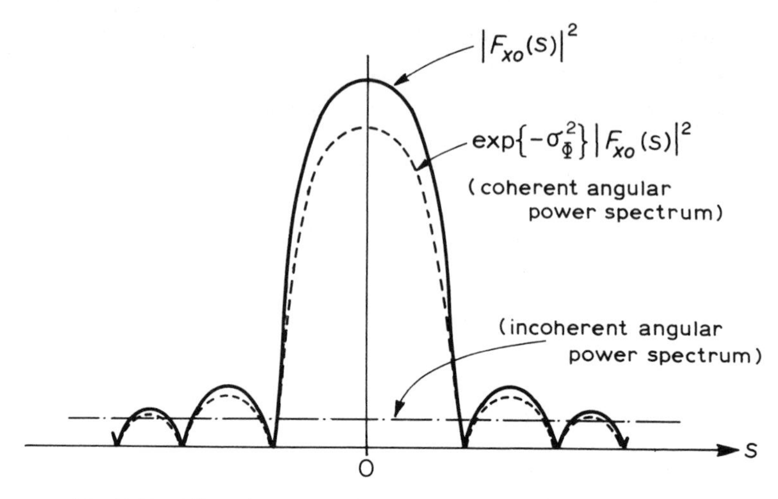

Fig. 7.14 – The coherent angular power spectrum for $\sigma_\Phi < 1$ radian.

The power 'lost' in this way reappears as incoherently scattered power. If the lateral scale size of the surface irregularities is significantly smaller than the overall aperture dimensions, then the angular distribution of the incoherent, scattered power will be much wider than the main beam of the ideal pattern, and could be taken as being constant out to the first few sidelobes at least, as assumed in Fig. 7.14. The effects of random profile irregularities are to reduce the gain and (on average) to increase the level of the sidelobes and to fill in the nulls in between.

The 'coherent gain', which is the gain calculated on the basis of the mean (coherent) radiated field of equation (7.100), is, by making use of equations (4.7) and (4.10),

$$G_{\text{coh}} = \exp\{-\sigma_\Phi^2\}\, G_o \tag{7.101}$$

where G_o is the gain of the antenna with no surface irregularities. G_{coh} slightly underestimates the actual gain of an antenna with reflector surface irregularities because it neglects the incoherent power in the direction $s = 0$. (If the lateral scale size of the irregularities is d, and the aperture is a circle of diameter D, then it can be shown that the on-axis incoherent power is of the order of $(\sigma_\Phi d/D)^2$ times the coherent, provided $\sigma_\Phi \ll 1$.) Hence it can be said that surface irregularities will produce a significant reduction in gain, that is, more than 4.3 dB, if

$$\sigma_\Phi \geqslant 1 \quad , \tag{7.102}$$

which is when (from equation 7.96)

$$\sigma_{\zeta_1} \geqslant \frac{\lambda}{4\pi} \approx 0.08\,\lambda \tag{7.103}$$

7.3 CASSEGRAIN REFLECTOR ANTENNAS

In the Cassegrain system a sub-reflector of hyperboloidal shape is used, as shown in Fig. 7.15 to redirect the radiation from the primary feed, which is now in the region of the apex of the main paraboloidal reflector, back on to the main reflector. In terms of rays, if the centre of radiation of the primary feed is at one of the foci F_1 of the hyperboloid then the ray reflected from any point P on the surface of the hyperboloidal sub-reflector appears to come from the other focus F_2. Then if F_2 is coincident with the focus of the paraboloidal main reflector then all rays reflected from it will emerge parallel to the main axis of the antenna.

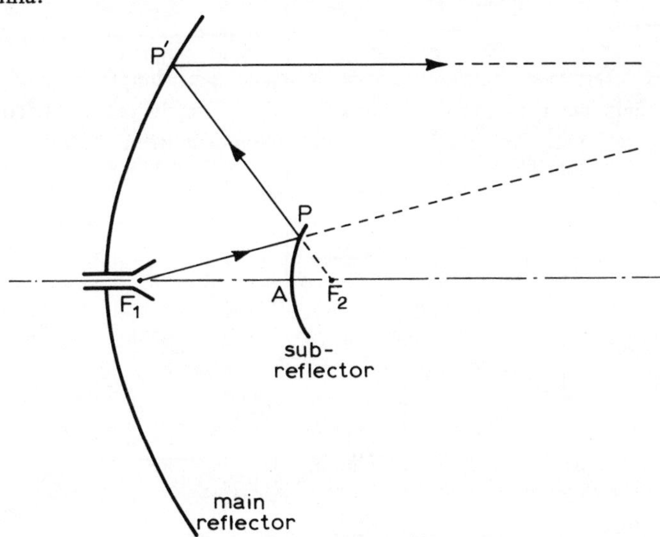

Fig. 7.15 – A typical ray path in the Cassegrain reflector antenna.

Most analyses of Cassegrain antennas are carried out, at least initially, using geometrical optics. Since the lateral dimensions of the primary feed are no more than one or two wavelengths, the sub-reflector can safely be assumed to be in the far zone of the feed. Hence geometrical optics can be used to describe the field incident on the sub-reflector. Now the lateral dimensions of the sub-reflector and its radius of curvature, are likely to be many wavelengths. Hence geometrical optics can again be used to describe the locally reflected field. But now the question must be asked as to whether the main reflector is in the far zone of the field reradiated by the sub-reflector. It usually is, as the following analysis shows.

If the primary field incident on the sub-reflector is circularly symmetric about the antenna axis, the equivalent aperture field on reflection will have the form

$$E_{ax}(\rho) = E_0(\rho)\exp\left\{-jk\,\frac{\rho^2}{2f_2}\right\} \tag{7.104}$$

over the circular projected aperture ○ and zero outside. This aperture field derives from the fact that the field reflected by the sub-reflector is locally a spherical wave centred on F_2, and $f_2 = AF_2$ is the focal length of the hyperboloid. The far field for this aperture field is given by the angular spectrum (in Cartesian form for ease of comparision later)

$$F_x(\alpha,\beta) = \frac{1}{\lambda^2} \iint_{\circ} E_o(x',y') \exp\left\{ -j\, \frac{k(x'^2 + y'^2)}{2f_2} \right\}$$

$$\exp\{jk(\alpha x' + \beta y')\}\, dx'dy' \quad . \tag{7.105}$$

In order to determine whether the main reflector is in the far zone of the sub-reflector's field we examine the same field in the Fresnel region. According to equation (5.84) then, the field at an axial distance z from the aperture is, on inserting equation (7.104),

$$E_x(x,y,z) = \frac{j}{\lambda z} \exp\left\{ -jk\left(z + \frac{x^2+y^2}{2z}\right)\right\}$$

$$\iint_{\circ} E_o(x',y') \exp\left\{ -j\,\frac{k}{2}\left(\frac{1}{z} + \frac{1}{f_2}\right)(x'^2 + y'^2)\right\}$$

$$\exp\left\{jk\left(\frac{x}{z}\,x' + \frac{y}{z}\,y'\right)\right\} dx'dy' \tag{7.106}$$

If $z \gg f_2$ this field has the same form as the far field corresponding to the angular spectrum of equation (7.105). Hence if the main reflector is at a distance which is large in comparison with the focal length of the hyperboloidal sub-reflector, which is usually the case in practice, then the main reflector can be considered to be in the far zone of the field from the sub-reflector.

A further simplifying notion is employed in the analysis of Cassegrain antennas, in which the combination primary feed/hyperboloidal sub-reflector/main paraboloidal reflector is replaced by the simpler combination of primary feed/equivalent paraboloidal reflector. Fig. 7.16 illustrates this. The ray from the feed at F_1 that strikes the sub-reflector at P is reflected from the actual main reflector at P$'$, and thence travels parallel to, and at a distance ρ from, the main axis. Instead, it can be imagined that the ray F_1 P continues beyond the sub-reflector in a straight line until it strikes an imaginary paraboloid reflector at P$''$ such that the ray is reflected parallel to, and at the same distance ρ from, the antenna axis. This imaginary paraboloid is known as the **equivalent paraboloid,** and has a focal length f_{eq} which can be calculated as shown below.

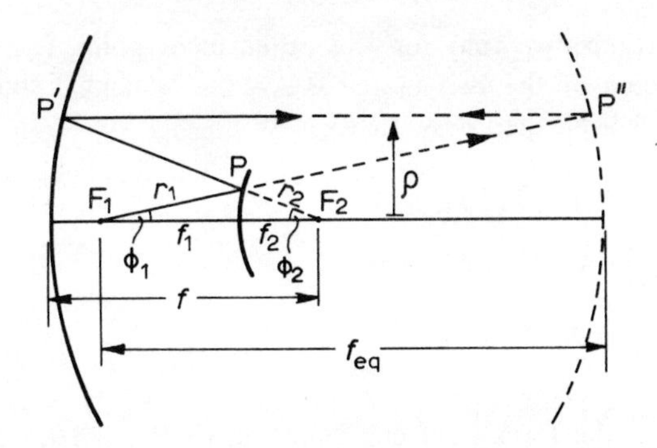

Fig. 7.16 – The Cassegrain and its 'equivalent paraboloid'.

Exercise

But first we need to review the properties of the hyperbola, whose surface of revolution about its axis is the hyperboloid. The two branches of the hyperbola are shown in Fig. 7.17; they are mirror images of each other in the ρ-axis. The hyperbola is defined as the locus of the point P (or P′) such that its distance from

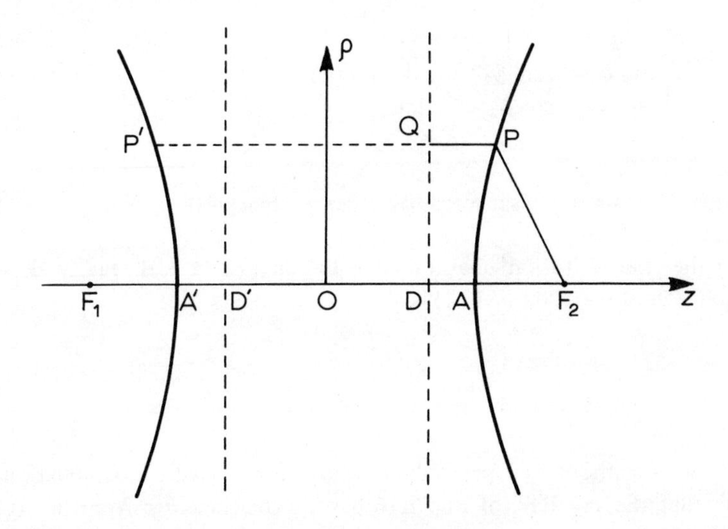

Fig. 7.17 – Geometry of the hyperbola.

the focal point F_2 bears a constant ratio ($e > 1$) to its distance PQ from the fixed line DQ, known as the **directrix**. Thus

$$\frac{F_2P}{PQ} = \frac{F_2A}{AD} = \frac{F_2P'}{P'Q} = e \quad , \tag{7.107}$$

and similar relationships hold for the other focal point F_1. If the distance between the apices of the two branches $AA' = 2a$, and if the focal distances $F_1A = f_1$ and $F_2A = f_2$, then show that

$$F_1P - F_2P = f_1 - f_2 = 2a \tag{7.108}$$

and that

$$f_1 + f_2 = 2ae \quad . \tag{7.109}$$

Returning to the Cassegrain configuration of Fig. 7.16, we may therefore write down the geometrical relationships

$$\begin{aligned} &r_1 - r_2 = f_1 - f_2 = 2a \\ &r_1 \sin \phi_1 - r_2 \sin \phi_2 = 0 \\ &r_1 \cos \phi_1 + r_2 \cos \phi_2 = f_1 + f_2 = 2ae \quad . \end{aligned} \tag{7.110}$$

In order that these three simultaneous equations in r_1 and r_2 have a non-trivial solution the determinant of the coefficients must be zero, that is,

$$\begin{vmatrix} 1 & -1 & 2a \\ \sin \phi_1 & -\sin \phi_2 & 0 \\ \cos \phi_1 & \cos \phi_2 & 2ae \end{vmatrix} = 0 \tag{7.111}$$

or

$$e(\sin \phi_1 - \sin \phi_2) + \sin \phi_1 \cos \phi_2 + \cos \phi_1 \sin \phi_2 = 0 \quad . \tag{7.112}$$

Using the 't substitution' method of equation (7.10) it is readily shown that the solution of equation (7.112) is

$$\tan \frac{\phi_2}{2} = \frac{e+1}{e-1} \tan \frac{\phi_1}{2} \tag{7.113}$$

Now, for the actual paraboloid, whose focal length is f, equation (7.11) shows that the ray P′P″ of Fig. 7.16 has a displacement ρ from the axis given by

$$\rho = 2f \tan \frac{\phi_2}{2} \quad . \tag{7.114}$$

But for the equivalent paraboloid the same displacement

$$\rho = 2f_{eq} \tan \frac{\phi_1}{2} \tag{7.115}$$

It then follows, by combining equations (7.113) to (7.115), that the focal length of the equivalent paraboloid is

$$f_{eq} = \frac{e+1}{e-1} f \quad . \tag{7.116}$$

In practice f_{eq} may be many times the actual focal length f.

Exercise
Establish the condition of equation (7.111) and its solution, equation (7.113).

7.4 OFFSET PARABOLOID-REFLECTOR ANTENNAS

In order to avoid aperture blocking effects, and radiation reflected back into the feed, the paraboloid reflector and Cassegrain designs considered in previous sections can be modified to employ offset sections of the main paraboloidal reflector producing a projected aperture which is not obstructed by either the feed or the sub-reflector. The simplest case is that of the offset paraboloid-reflector antenna of Fig. 7.18.

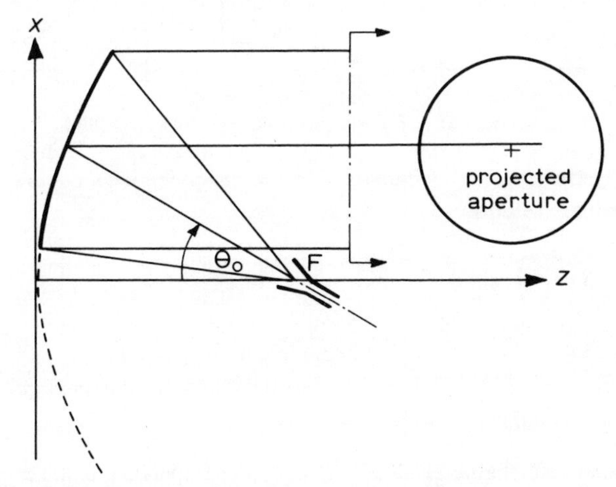

Fig. 7.18 – A primary fed offset paraboloid-reflector antenna.

If the feed is x-polarized (that is, has no y-component of electric field in its aperture) the field incident on the reflector surface will also be x-polarized, even when the axis of the feed is tilted back through the offset angle θ_0. Thus equation (7.17) still applies, with the angular spectrum modified to allow for the change in pointing. With the feed at the focus F of the paraboloidal surface, the projected aperture field will acquire a constant phase and its amplitude will depend on the feed pattern and the distance of the feed to each point of the reflector, as before.

There is a difference, however, in the polarization of the projected aperture field compared to the axisymmetrically fed case. This can be seen by superimposing the offset projected aperture onto the field pattern of Fig. 7.6. The resulting electric-field pattern is shown in Fig. 7.19 where it is also resolved schematically into its x- and y-components. From this it can be seen that while the x-component is approximately uniform (assuming uniform illumination) the y-component is bunched up into two regions approximately $3D/4$ apart, where they are also oppositely directed. One can therefore argue physically that the cross-polar far field diffracted from this projected aperture will be zero on axis and in the (vertical) plane of symmetry, due to cancellation of the opposing y-components. But in the (horizontal) plane of asymmetry the cross-polar field will have a lobe structure, which can be crudely estimated as follows.

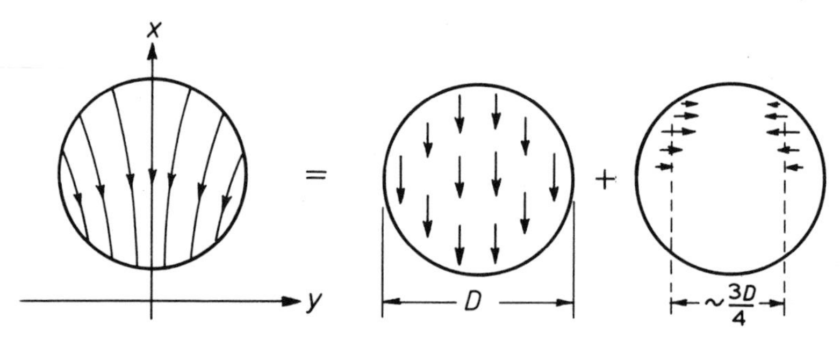

Fig. 7 19 – Off-set projected aperture field lines

If θ_c is the angle at which the first cross-polar lobe occurs, it will satisfy the approximate relation

$$k\left(\frac{3D}{4}\right)\sin\theta_c = \pi$$

or

$$\theta_c \approx \frac{2\lambda}{3D}\,, \tag{7.117}$$

assuming that the diameter D of the projected aperture is much larger than λ. Comparing this with the copolar pattern for a uniformly illuminated circular aperture of diameter D, equation (7.32) and Fig. 7.8 show that

$$\frac{2J_1(u)}{u} = \tfrac{1}{2}\;;\quad u = \frac{\pi D\sin\theta}{\lambda}$$

when $u \approx 2$, or

$$\theta \approx \frac{2\lambda}{\pi D} \tag{7.118}$$

Thus, comparing equations (7.117) and (7.118), the first lobe of the cross-polar pattern occurs at about the -6 dB level of the co-polar pattern, for a linearly x-polarized feed. The same arguments apply to a purely y-polarized feed, simply by rotating Fig. 7.6 through 90°. It is also clear, physically, that the overall magnitude of the cross-polar patterns will be reduced the larger the f/D ratio and the smaller the offset angle θ_0.

Of course, more detailed calculations, based on appropriate modifications of the aperture fields derived in section 7.2 for the axisymmetrically-fed paraboloid reflector, can be performed and yield useful results. In particular, it can be shown that a circularly polarized feed has no cross-polarized component in the far field, but produces a beam squint across the plane of symmetry, the direction of the squint depending on the sense of polarization. (See Rudge and Adatia (1978).)

7.5 THE HORN-PARABOLOID ANTENNA

If a section of a paraboloidal surface is attached to an electromagnetic horn to form a cowl (that is, a monk's hood), as shown in Fig. 7.20, with the phase centre of the horn at the focus of the paraboloid, the spherical wave emerging from the mouth of the horn is transformed into a projected aperture field whose wavefront is plane. The paraboloidal section therefore acts as a lens, but without giving rise to the partial reflections that go with a lens placed across the mouth of the horn (as in Fig. 1.4). The price paid is the inevitable occurrence of cross-polarized components in the projected aperture field.

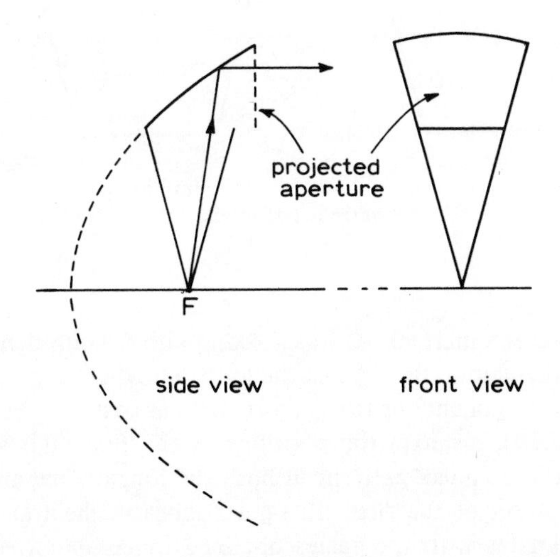

Fig. 7.20 – Schematic construction of the horn-paraboloid antenna.

Assuming that the horn is pyramidal (see section 7.1) and fed by a rectangular waveguide supporting the TE_{01} mode, the amplitude distribution of this mode will be preserved throughout the length of the horn, and so the amplitude of the projected aperture field will be essentially of the same form. Allowance must be made for the non-rectangular form of the projected aperture, which is achieved in the usual way by projecting the conical rays emerging from the horn forward on to the paraboloidal surface and assuming that they then emerge as parallel-ray tubes. This construction implies that there will be an amplitude taper in the direction of the farthest edge superimposed on the basic TE_{01} distribution.

This same geometrical construction provides a description of the polarization of the electric field over the projected aperture. Fig. 7.21 shows the two cases of (a) when the electric field in the feeder waveguide is parallel to the plane of symmetry of the antenna, and is referred to as **longitudinal polarization,** and (b) when the electric field is **transverse** to this plane. It can be seen that the fields are similar in character to those found in the case of the offset-reflector antenna. The cross-polarized components are oppositely directed on either side of the plane symmetry, and increase in magnitude with increasing distance from this plane. Thus, there will be negligible cross-polarization in the boresight direction and in the (longitudinal) plane of symmetry. But cross-polarization will appear in the transverse plane, with the first lobe occurring at roughly the same position as that estimated for the offset-reflector antenna.

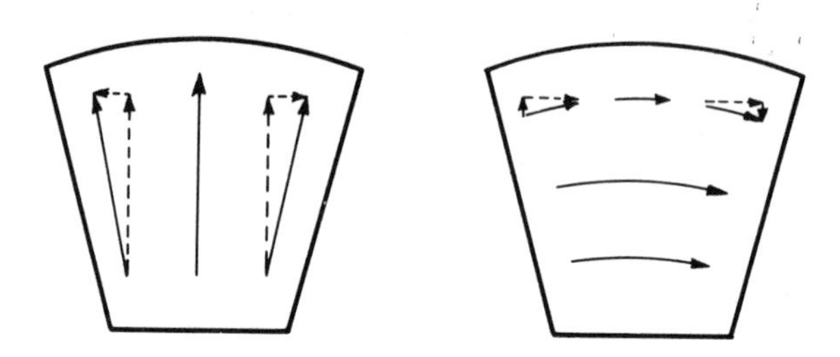

Fig. 7.21 – Projected aperture fields for (a) longitudinal polarization and (b) transverse polarization.

Exercise

By making drastic approximations, such as working with an equivalent rectangular aperture, ignoring the amplitude taper in the longitudinal direction, and treating the cross-polarized components of the aperture field as occurring in two distinct bunches (cf. Fig. 7.19), estimate the positions of the first nulls and the first sidelobe levels of the co-polar patterns in both the longitudinal and transverse planes, and the position of the first cross-polar lobes in the transverse plane. Compare these estimates with the values obtained by computer calculation in the original reference (Crawford, Hogg and Hunt, 1961).

Apart from its aperture, the horn-paraboloid antenna is entirely enclosed by metal. It is therefore mechanically very robust, though heavy. An electrical advantage of its being enclosed is that the radiation in most directions other than in the region of boresight is minimal, as compared to open structures such as horn-fed paraboloids and Cassegrains. This ensures good isolation in communication applications. In particular, as a ground-station antenna for a satellite communication link, environmental noise is kept to a minimum. (Noise due to the Earth's environment and to known cosmic sources was calculated with such accuracy by Bell Laboratories engineers for this type of antenna that they were able to identify with certainty the presence of a 4°K background noise of cosmic microwave radiation (Ohm (1961) and Penzias and Wilson (1965)).

One direction in which spillover is not negligible, although still not large, is by way of diffraction at the furthest edge of the paraboloidal cowl. Knife-edge diffraction theory (see sections 5.2 and 5.5.2) can be used to predict the amplitude of the spillover lobe fairly precisely, provided allowance is made for the curvature of the diffracting edge. Spillover will be much less pronounced, of course, when the polarization is transverse rather than longitudinal. This is due to the cosine amplitude taper towards the edge in the case of transverse polarization.

Exercise

Reformulate the three-dimensional knife-edge analysis of section 5.5.2 for it to apply to diffraction by a curved edge. This can be done by setting $h = y^2/(2R_e)$ where R_e is the radius of curvature of the edge. Estimate the effect of edge curvature on the diffracted field.

CHAPTER 8

Radiation from non-planar apertures

The analysis has so far been restricted to radiation from planar apertures. In this final chapter this restriction will be lifted, and a formula obtained giving the far-field vector pattern function in terms of fields specified over any arbitrary, closed surface. This formula will be seen to have useful applications in the realm of near-field antenna measurements.† It also gives a concise overall view of the antenna analysis problem, and in particular reveals the reason why the planar aperture approach is so often favoured, not only for its simplicity but also for its precision.

First a general result, termed the near-field to far-field formula, will be derived. Its basis is the reciprocity theorem and many of the details of its derivation are those that have already been employed in Chapter 4. The problem, stated pictorially in Fig. 8.1, is: given the field $\mathbf{E}_o, \mathbf{H}_o$ specified over the closed surface S, what is the far-field vector pattern function $\mathbf{e}(\mathbf{u})$ of the resulting radiation.

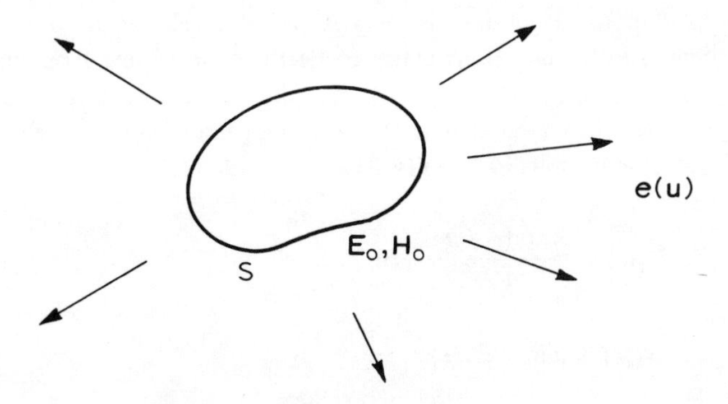

Fig. 8.1 – Radiation from the non-planar aperture S.

† Chappas (1974) has examined some of these applications.

8.1 DERIVATION OF THE NEAR-FIELD TO FAR-FIELD FORMULA

Apply the Lorentz reciprocity theorem (see appendix B.2.8) in the form

$$\int_S (\mathbf{E}_1 \times \mathbf{H}_2 - \mathbf{E}_2 \times \mathbf{H}_1) \,.\, \mathbf{n}\, da = \int_{S_1} (\mathbf{E}_1 \times \mathbf{H}_2 - \mathbf{E}_2 \times \mathbf{H}_1) \,.\, \mathbf{u}_r da \tag{8.1}$$

to the source-free volume V (see Fig. 8.2) lying between the arbitrary-shaped closed surface S and S_1, which is the surface of a sphere of large radius R.

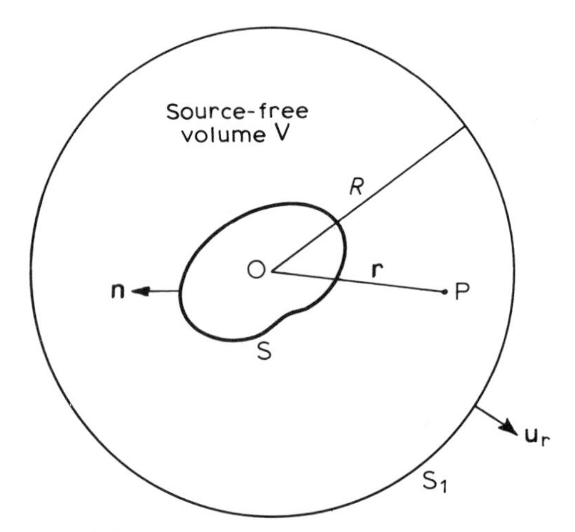

Fig. 8.2 – Geometry of the source-free volume V bounded by S and by a sphere S_1 of large radius R.

The outward normal at all points on S is the unit vector $\mathbf{n}$, and similarly $\mathbf{u}_r$ is the outward normal for the sphere S_1. The fields $\mathbf{E}_1$, $\mathbf{H}_1$ and $\mathbf{E}_2$, $\mathbf{H}_2$ are two solutions of Maxwells equations at frequency ω within V, specified at the general field point P whose vector position is $\mathbf{r}$ relative to the centre of the sphere. Specifically, $\mathbf{E}_1$, $\mathbf{H}_1$ is taken to be the field arising from the prescribed field $\mathbf{E}_0$, $\mathbf{H}_0$ on S, and assumed to be radiating wholly away from S into V. Hence, at sufficiently large distances $\mathbf{r}$ from O, this radiated field can be expressed in the form of equations (4.1) to (4.3) as

$$\mathbf{E}_1(\mathbf{r}) = \frac{\exp(-jkr)}{kr}\; \mathbf{e}(\mathbf{u}_r) \tag{8.2}$$

$$\mathbf{H}_1(\mathbf{r}) = Y\mathbf{u}_r \times \mathbf{E}_1(\mathbf{r}) \tag{8.3}$$

with

$$\mathbf{u}_r \,.\, \mathbf{e}(\mathbf{u}_r) = 0 \tag{8.4}$$

to an increasingly good approximation as $kr \to \infty$. $Y = Z^{-1}$ is the intrinsic admittance of free space, and $\mathbf{e}(\mathbf{u}_r)$ is the far-field vector pattern function of the radiation.

The second field $\mathbf{E}_2$, $\mathbf{H}_2$, will be taken to be the combination of a plane wave $\mathbf{e}_i$ incident on S from the direction $\mathbf{u}_i$ together with the resulting field scattered by S, namely

$$\mathbf{E}_2(\mathbf{r}) = \mathbf{e}_i \exp(jk\mathbf{u}_i \cdot \mathbf{r}) + \mathbf{E}_s(\mathbf{r}) \tag{8.5}$$

$$\mathbf{H}_2(\mathbf{r}) = -Y\mathbf{u}_i \times \mathbf{e}_i \exp(jk\mathbf{u}_i \cdot \mathbf{r}) + \mathbf{H}_s(\mathbf{r}) \tag{8.6}$$

with

$$\mathbf{u}_i \cdot \mathbf{e}_i = 0 \ . \tag{8.7}$$

For sufficiently large r the scattered field $\mathbf{E}_s(\mathbf{r})$, $\mathbf{H}_s(\mathbf{r})$ will have the same form as the radiated field of equations (8.2) to (8.4) with an appropriate far-field vector pattern function $\mathbf{e}_s(\mathbf{r})$.

Referring to the proof of the antenna reciprocity theorem in section 4.2.1, it is clear from equations (4.43) and (4.44) that the above scattered field does not contribute to the integral on the right-hand side of equation (8.1). As regards the contribution of the plane-wave part of the field $\mathbf{E}_2$, $\mathbf{H}_2$ to the reciprocity integral, it again follows from the reciprocity theorem proof (see equations (4.38) to (4.41)) that

$$\int_{S_1} (\mathbf{E}_1 \times \mathbf{H}_2 - \mathbf{E}_2 \times \mathbf{H}_1) \cdot \mathbf{u}_r \, da = \frac{j\lambda^2 Y}{\pi} \mathbf{e}_i \cdot \mathbf{e}(\mathbf{u}_i) \tag{8.8}$$

which becomes increasingly exact as $kR \to \infty$, and $\mathbf{e}(\mathbf{u}_i)$ precisely represents the radiated field.

Finally, equations (8.1) and (8.8) together yield the desired near-field to far-field formula, which is

$$\mathbf{e}_i \cdot \mathbf{e}(\mathbf{u}_i) = \frac{\pi}{j\lambda^2 Y} \int_S [\mathbf{E}_1(\mathbf{r}_o) \times \mathbf{H}_2(\mathbf{r}_o) - \mathbf{E}_2(\mathbf{r}_o) \times \mathbf{H}_1(\mathbf{r}_o)] \cdot \mathbf{n} \, da \tag{8.9}$$

in which points on the surface S are given by $\mathbf{r} = \mathbf{r}_o$. Equation (8.9) will be used to evaluate the far-field radiation pattern $\mathbf{e}(\mathbf{u})$ arising from the prescribed fields $\mathbf{E}_1(\mathbf{r}_o)$, $\mathbf{H}_1(\mathbf{r}_o)$ on S (namely, $\mathbf{E}_o$, $\mathbf{H}_o$ of Fig. 8.1). It essentially expresses the result of probing $\mathbf{e}(\mathbf{u})$ with a plane wave $\mathbf{e}_i$ arriving from the direction $\mathbf{u}_i$. Different forms of the result will be obtained subsequently, their differences arising from whatever assumptions are made concerning the boundary conditions to be satisfied on S by the field $\mathbf{E}_2$, $\mathbf{H}_2$. But the general form (8.9) shows, using the scalar triple-product indentity

$$(\mathbf{E}_1 \times \mathbf{H}_2) \,.\, \mathbf{n} = \mathbf{E}_1 \,.\, (\mathbf{H}_2 \times \mathbf{n}) = \mathbf{H}_2 \,.\, (\mathbf{n} \times \mathbf{E}_1) \tag{8.10}$$

and similarly for $\mathbf{E}_2$ and $\mathbf{H}_1$ on S, that the integration over S on the right-hand side of equation (8.9) involves the tangential components only of the two sets of fields.

8.2 THE KIRCHHOFF-HUYGENS FORMULA

If the region within S is chosen to have the same constitutive parameters ϵ, μ as the uniform exterior, then the scattered-field contribution to $\mathbf{E}_2$, $\mathbf{H}_2$ will be zero and

$$\mathbf{E}_2(\mathbf{r}) = \mathbf{e}_\mathrm{i} \exp(jk\mathbf{u} \,.\, \mathbf{r}) \tag{8.11}$$

$$\mathbf{H}_2(\mathbf{r}) = -\,Y\mathbf{u} \times \mathbf{e}_\mathrm{i} \exp(jk\mathbf{u} \,.\, \mathbf{r}) \tag{8.12}$$

where the subscript i on $\mathbf{u}$ has been dropped. Then, from equation (8.9),

$$\mathbf{e}_\mathrm{i} \,.\, \mathbf{e}(\mathbf{u}) = \frac{\pi}{j\lambda^2 Y} \int_S [-Y\mathbf{E}_\mathrm{o} \times (\mathbf{u} \times \mathbf{e}_\mathrm{i}) \exp(jk\mathbf{u} \,.\, \mathbf{r}_\mathrm{o}) - \mathbf{e}_\mathrm{i} \times \mathbf{H}_\mathrm{o} \exp(jk\mathbf{u} \,.\, \mathbf{r}_\mathrm{o})] \,.\, \mathbf{n}\, \mathrm{d}a. \tag{8.13}$$

Using the identities in equation (8.10) it follows that

$$\mathbf{E}_\mathrm{o} \times (\mathbf{u} \times \mathbf{e}_\mathrm{i}) \,.\, \mathbf{n} = -\mathbf{e}_\mathrm{i} \,.\, \mathbf{u} \times (\mathbf{n} \times \mathbf{E}_\mathrm{o})$$

and that

$$\mathbf{e}_\mathrm{i} \times \mathbf{H}_\mathrm{o} \,.\, \mathbf{n} = -\mathbf{e}_\mathrm{i} \,.\, \mathbf{n} \times \mathbf{H}_\mathrm{o} \quad .$$

Equation (8.13) thus becomes

$$\mathbf{e}_\mathrm{i} \,.\, \mathbf{e}(\mathbf{u}) = \frac{\pi}{j\lambda^2} \int_S [\mathbf{e}_\mathrm{i} \,.\, \mathbf{u} \times (\mathbf{n} \times \mathbf{E}_\mathrm{o}) + Z\, \mathbf{e}_\mathrm{i} \,.\, \mathbf{n} \times \mathbf{H}_\mathrm{o}] \exp(jk\mathbf{u} \,.\, \mathbf{r}_\mathrm{o})\, \mathrm{d}a \quad . \tag{8.14}$$

The above equation holds for any vector $\mathbf{e}_i$ normal to $\mathbf{u}$. The vector $\mathbf{e}(\mathbf{u})$ is also normal to $\mathbf{u}$ and it can be shown by straightforward vector algebra that

$$\mathbf{e}(\mathbf{u}) = \frac{\pi}{j\lambda^2} \int_S [\mathbf{u} \times (\mathbf{n} \times \mathbf{E}_o) + Z\, \mathbf{u} \times \{(\mathbf{n} \times \mathbf{H}_o) \times \mathbf{u}\}] \exp(jk\mathbf{u} \cdot \mathbf{r}_o)\, da \tag{8.15}$$

which gives the far-field vector pattern function $\mathbf{e}(\mathbf{u})$ as an integral of the field $\mathbf{E}_o$, $\mathbf{H}_o$ over the closed surface S. Note that both the electric and magnetic fields must be specfied.

The two terms $(\mathbf{n} \times \mathbf{H}_o)$ and $(\mathbf{n} \times \mathbf{E}_o)$ in the integrand of equation (8.15) are respectively the tangential magnetic and electric fields on S. They correspond physically to equivalent electric and magnetic surface currents

$$\mathbf{J}_s = \mathbf{n} \times \mathbf{H}_o \quad \text{and} \quad K_s = -\mathbf{n} \times \mathbf{E}_o \tag{8.16}$$

induced on the surface of S. Radiation from these currents is the basis of Schelkunoff's (1943) equivalence principle for the calculation of radiation from aperture fields, and forms a direct mathematical statement of Huygens' principle.

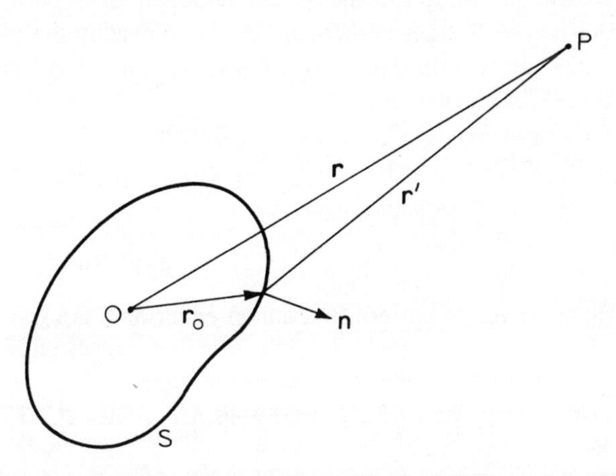

Fig. 8.3 – Geometry for the statement of the Kirchhoff-Huygens formula in its conventional form.

Equation (8.15) is identical to the Kirchhoff-Huygens formula of classical antenna theory, as we now demonstrate. To quote Silver (1949, p. 107), and referring to Fig. 8.3, the conventional formula for the electric field at P is

$$\mathbf{E}_p = \frac{1}{4\pi} \int_S [-j\omega\mu\, (\mathbf{n} \times \mathbf{H})\, \psi + (\mathbf{n} \times \mathbf{E}) \times \nabla\psi + (\mathbf{n} \cdot \mathbf{E})\nabla\psi]\, da \tag{8.17}$$

in which

$$\psi = \frac{\exp(-jkr')}{r'} \tag{8.18}$$

with r' measured from P to a point on the surface S. Taking its gradient,

$$\nabla \psi = -(jk + \frac{1}{r'}) \frac{\exp(-jkr')}{r'} \mathbf{u}_r'$$

As the observation point P approaches infinity the gradient of ψ tends to

$$\nabla \psi = \frac{jk}{r} \exp [jk(\mathbf{u} \cdot \mathbf{r}_o - r)] \, \mathbf{u} \tag{8.19}$$

with the result that the far-field becomes

$$\mathbf{E}_p = \frac{\exp(-jkr)}{4\pi r} \int_S \left[-j\omega\mu \, (\mathbf{n} \times \mathbf{H}) + jk \, \{(\mathbf{n} \times \mathbf{E}) \times \mathbf{u} + (\mathbf{n} \cdot \mathbf{E}) \, \mathbf{u}\} \right] \exp (jk\mathbf{u} \cdot \mathbf{r}_o) \, da \, . \tag{8.20}$$

The term $(\mathbf{n} \cdot \mathbf{E})\mathbf{u}$ contributes to the radial component at P and it may be shown, by application of the divergence theorem, to cancel exactly the radial component arising from the $\mathbf{n} \times \mathbf{H}$ term. Since the total radial component is zero in the far field, equation (8.20) may be rewritten

$$\mathbf{E}_p = \frac{\exp(-jkr)}{kr} \cdot \frac{\pi}{j\lambda^2} \int_S [\mathbf{u} \times (\mathbf{n} \times \mathbf{E}) + Z \, \mathbf{u} \times \{(\mathbf{n} \times \mathbf{H}) \times \mathbf{u}\}] \exp (jk\mathbf{u} \cdot \mathbf{r}_o) \, da \tag{8.21}$$

which has the same vector pattern function as equation (8.15).

8.3 RELATIONSHIP BETWEEN FORMULAS FOR SCATTERING AND RADIATION

Suppose now that the surface S is covered by a perfect conductor. The boundary condition for the total field $\mathbf{E}_2$ on S is then

$$\mathbf{n} \times \mathbf{E}_2(\mathbf{r}_o) = 0 \tag{8.22}$$

and the second term in the integrand of equation (8.9) consequently vanishes, because

$$\mathbf{E}_2(\mathbf{r}_o) \times \mathbf{H}_1(\mathbf{r}_o) \cdot \mathbf{n} = \mathbf{H}_1(\mathbf{r}_o) \cdot (\mathbf{n} \times \mathbf{E}_2(\mathbf{r}_o)) = 0 \ . \tag{8.23}$$

The field $\mathbf{H}_2(\mathbf{r}_o)$ is now the total magnetic field on a conducting obstacle with surface S, when illuminated by a plane wave incident from the direction **u**. Denote this total field by $\mathbf{H}_2^c(\mathbf{r}_o\,;\mathbf{u})$, and equation (8.9) becomes

$$\mathbf{e}_i \cdot \mathbf{e}(\mathbf{u}) = \frac{\pi}{j\lambda^2 Y} \int_S [\mathbf{E}_1(\mathbf{r}_o) \times \mathbf{H}_2^c(\mathbf{r}_o\,;\mathbf{u})] \cdot \mathbf{n}\, da \quad . \tag{8.24}$$

This equation expresses the radiation field **e(u)** as an integral depending only on the tangential component of the electric field $\mathbf{n} \times \mathbf{E}_1(\mathbf{r}_o)$ on S. The 'kernel' $\mathbf{H}_2^c(\mathbf{r}_o\,;\mathbf{u})$ is derived from the associated problem of scattering by a conducting obstacle with surface S.

It is to be expected that formula (8.24) will lead to a particularly simple result when the conducting surface S is an infinite plane. The geometry of section 8.1 needs only slight modification, in that S_1 should now be a hemisphere of infinite radius. The same formulas still apply, with **e(u)** restricted to giving the far field in the half-space on the side of S containing the incident, probing plane wave $\mathbf{e}_i$.

Define S as the plane, depicted in Fig. 8.4, for which

$$\mathbf{n} \cdot \mathbf{r}_o = 0 \tag{8.25}$$

with **n** constant. Thus $\mathbf{r}_o$ defines any point in the plane. Take the direction of the incident plane wave to be $\mathbf{u}_i = -\mathbf{u}$ and its vector amplitude to be $\mathbf{e}_i$, referred to O as phase reference.

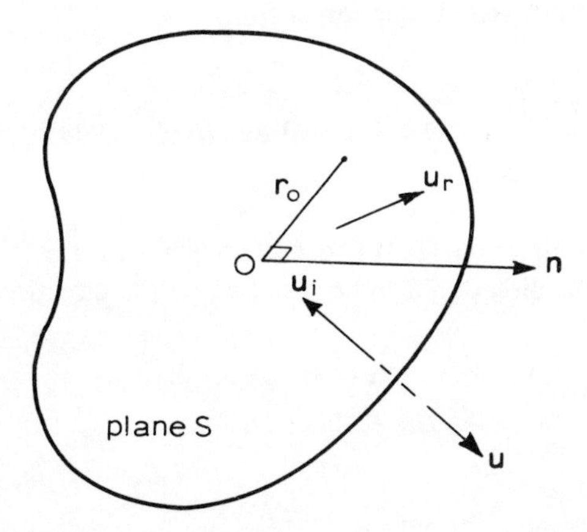

Fig. 8.4 – Position vector $\mathbf{r}_o$ in the plane S, defined by the constant **n**, such that $\mathbf{n} \cdot \mathbf{r}_o = 0$.

The results of section 6.1 give the total magnetic field at the conductor surface as

$$\mathbf{H}_2^c(\mathbf{r}_o\,;\mathbf{u}) = 2Y\,[(\mathbf{n}\,.\,\mathbf{u}\times\mathbf{e}_i)\,\mathbf{n} - \mathbf{u}\times\mathbf{e}_i]\;\exp\,(jk\mathbf{u}\,.\,\mathbf{r}_o)\quad . \tag{8.26}$$

Now, from the identities (8.10), equation (8.24) can be rewritten as

$$\mathbf{e}_i\,.\,\mathbf{e}(\mathbf{u}) = \frac{\pi}{j\lambda^2 Y}\int_S [\mathbf{E}_1(\mathbf{r}_o)\,.\,\mathbf{H}_2^c(\mathbf{r}_o\,,\mathbf{u})\times\mathbf{n}]\;da \tag{8.27}$$

and from equation (8.26)

$$\mathbf{H}_2^c(\mathbf{r}_o\,;\mathbf{u})\times\mathbf{n} = 2Y\mathbf{n}\times(\mathbf{u}\times\mathbf{e}_i)\,\exp\,(jk\mathbf{u}\,.\,\mathbf{r}_o) \tag{8.28}$$

Further use of (8.10) gives that

$$\mathbf{E}_1(\mathbf{r}_o)\,.\,\mathbf{n}\times(\mathbf{u}\times\mathbf{e}_i) = \{\mathbf{E}_1(\mathbf{r}_o)\times\mathbf{n}\}\times\mathbf{u}\,.\,\mathbf{e}_i \tag{8.29}$$

and hence from (8.27) to (8.29) that

$$\mathbf{e}_i\,.\,\mathbf{e}(\mathbf{u}) = j\,\frac{2\pi}{\lambda^2}\int_S \mathbf{u}\times\{\mathbf{E}_1(\mathbf{r}_o)\times\mathbf{n}\}\,.\,\mathbf{e}_i\,\exp\,(jk\mathbf{u}\,.\,\mathbf{r}_o)\;da\quad . \tag{8.30}$$

Then, finally, since $\mathbf{e}_i$ and $\mathbf{e}(\mathbf{u})$ are co-planar and the polarization of $\mathbf{e}_i$ is arbitrary, the far-field vector pattern function is

$$\mathbf{e}(\mathbf{u}) = j\,\frac{2\pi}{\lambda^2}\int_S \mathbf{u}\times\{\mathbf{E}_1(\mathbf{r}_o)\times\,\mathbf{n}\}\,\exp\,(jk\mathbf{u}\,.\,\mathbf{r}_o)\;da\quad . \tag{8.31}$$

Converting equation (8.31) to rectangular coordinates, taking S to be the x-y plane and the direction $\mathbf{u}$ to be specified by the direction cosines (α,β,γ),

$$\mathbf{u} = \mathbf{u}_x\alpha + \mathbf{u}_y\beta + \mathbf{u}_z\gamma\quad ,$$

$$\mathbf{n} = \mathbf{u}_z,$$

and

$$\mathbf{r}_o = \mathbf{u}_x x + \mathbf{u}_y y\quad ,$$

the vector pattern function is

$$\mathbf{e}(\alpha,\beta) = j\,2\pi\,\frac{1}{\lambda^2}\int_{-\infty}^{\infty}\int_{-\infty}^{\infty}\left\{E_{1x}(x,y)\,[\mathbf{u}_x\gamma - \mathbf{u}_z\alpha] + E_{1y}(x,y)\,[\mathbf{u}_y\gamma - \mathbf{u}_z\beta]\right\}$$

$$\exp\,[jk(\alpha x + \beta y)]\;\mathrm{d}x\,\mathrm{d}y \qquad (8.32)$$

$$= j2\pi\left\{F_x(\alpha,\beta)\,[\mathbf{u}_x\gamma - \mathbf{u}_z\alpha] + F_y(\alpha,\beta)\,[\mathbf{u}_y\gamma - \mathbf{u}_z\beta]\right\}\;, \qquad (8.33)$$

where

$$F_x(\alpha,\beta) = \frac{1}{\lambda^2}\int_{-\infty}^{\infty}\int_{-\infty}^{\infty} E_{1x}(x,y)\exp\{jk(\alpha x + \beta y)\}\,\mathrm{d}x\,\mathrm{d}y \qquad (8.34)$$

and

$$F_y(\alpha,\beta) = \frac{1}{\lambda^2}\int_{-\infty}^{\infty}\int_{-\infty}^{\infty} E_{1y}(x,y)\exp\{jk(\alpha x + \beta y)\}\,\mathrm{d}x\,\mathrm{d}y \qquad (8.35)$$

are the angular plane-wave spectra corresponding to the x- and y- components of the tangential electric field specified over the aperture (x-y) plane. This is precisely the form already given in section 3.5.1 for the vector far-field radiating into the region $z \geqslant 0$.

When the surface S is not an infinite plane the kernel $\mathbf{H}_2{}^{\mathrm{c}}(\mathbf{r}_o\,;\mathbf{u})$ is far more complicated. However, solutions to the associated scattering problem are available, or can be obtained numerically to any desired degree of accuracy. Hence equation (8.24) can be used as a basis for so-called near-field antenna measurements, in which the tangential electric field is recorded over a closed surface of observation S.

The formula for the complementary case, when the tangential magnetic field on S is recorded, is obtained by supposing S to be a perfect magnetic conductor. The boundary condition is

$$\mathbf{n} \times \mathbf{H}_2(\mathbf{r}_o) = 0\;\;. \qquad (8.36)$$

If the corresponding total electric field on the surface of the perfect magnetic conductor S, due to the plane wave $\mathbf{e}_i$ incident from the direction $\mathbf{u}$, is $\mathbf{E}_2{}^{\mathrm{m}}(\mathbf{r}_o;\mathbf{u})$ equation (8.9) becomes

$$\mathbf{e}_i \,.\, \mathbf{e}(\mathbf{u}) = \frac{\pi}{j\lambda^2 Y}\int_S [\mathbf{H}_1(\mathbf{r}_o) \times \mathbf{E}_2{}^{\mathrm{m}}(\mathbf{r}_o\,;\mathbf{u})]\;.\,\mathbf{n}\;\mathrm{d}a\;\;. \qquad (8.37)$$

This result can conveniently be expressed in terms of the previous kernel $\mathbf{H}_2^c(\mathbf{r}_o\,;\mathbf{u})$ in the following way.

In the case when S was a perfect electric conductor, $\mathbf{E}_2(r)$ and $\mathbf{H}_2(r)$ were the total fields in V composed of the plane wave $\mathbf{e}_i$ incident from the direction $\mathbf{u}$ and the resulting scattered field. The boundary conditions were

$$\mathbf{n} \times \mathbf{E}_2(\mathbf{r}_o) = 0 \quad (\text{on S}) \tag{8.38}$$

and at infinity (apart from the far scattered field which did not contribute)

$$\mathbf{E}_2(\mathbf{r}) \to \mathbf{e}_i \exp(jk\,\mathbf{u}\,.\,\mathbf{r}) \quad (\text{as } r \to \infty) \tag{8.39}$$

The duality principle (Appendix B.2.7) provides a second solution, which also satisfies Maxwell's equations in V, namely

$$\mathbf{E}_2{}'(\mathbf{r}) = Z\,\mathbf{H}_2(\mathbf{r}) \tag{8.40}$$

and

$$\mathbf{H}_2{}'(\mathbf{r}) = -Y\mathbf{E}_2(\mathbf{r}) \tag{8.41}$$

which satisfy the boundary conditions

$$\mathbf{n} \times \mathbf{H}_2'(\mathbf{r}) = 0 \quad (\text{on S}) \tag{8.42}$$

and (essentially)

$$\mathbf{E}_2'(\mathbf{r}) \to -\,\mathbf{u} \times \mathbf{e}_i \exp(jk\,\mathbf{u}\,.\,\mathbf{r}) \quad (\text{as } r \to \infty) \quad . \tag{8.43}$$

Thus $\mathbf{E}_2'(\mathbf{r})$ and $\mathbf{H}_2'(\mathbf{r})$, so defined, are the solution for the scattering problem of a plane wave $-\mathbf{u} \times \mathbf{e}_i$ incident from the direction $\mathbf{u}$ on S made of perfect magnetic conducting material. The total surface electric field is then

$$\mathbf{E}_2'(\mathbf{r}_o) = Z\,\mathbf{H}_2^c(\mathbf{r}_o\,;\mathbf{u}) = \mathbf{E}_2^m(\mathbf{r}_o\,;\mathbf{u}) \tag{8.44}$$

and so, by substituting (8.44) into (8.37),

$$(\mathbf{e}_i \times \mathbf{u})\,.\,\mathbf{e}(\mathbf{u}) = \frac{\pi}{j\lambda^2 Y^2} \int_S [\mathbf{H}_1(\mathbf{r}_o) \times \mathbf{H}_2^c(\mathbf{r}_o\,;\mathbf{u})]\,.\,\mathbf{n}\,da \quad . \tag{8.45}$$

Then, since from (8.10),

$$(\mathbf{e}_i \times \mathbf{u})\,.\,\mathbf{e}(\mathbf{u}) = \mathbf{e}_i\,.\,[\mathbf{u} \times \mathbf{e}(\mathbf{u})] = Z\,\mathbf{e}_i\,.\,\mathbf{h}(\mathbf{u}) \tag{8.46}$$

where $\mathbf{h}(\mathbf{u})$ is the vector magnetic far-field pattern function, we have finally that

$$\mathbf{e}_i \cdot \mathbf{h}(\mathbf{u}) = \frac{\pi}{j\lambda^2 Y} \int_S [\mathbf{H}_1(\mathbf{r}_o) \times \mathbf{H}_2^c(\mathbf{r}_o; \mathbf{u})] \cdot \mathbf{n}\, da \quad . \tag{8.47}$$

Thus a solution for scattering by a perfectly conducting body S provides the essential data from which the far-field vector pattern function can be calculated from a knowledge of either the tangential electric field or the tangential magnetic field over the closed surface S.

Appendix A

Transmission lines, plane waves, and simple waveguides

Since the plane wave is the basis of our presentation of diffraction theory, the properties of plane waves will be developed here. Instead of the usual formal development from Maxwell's equations, which can be found in many textbooks on electromagnetic waves, we have chosen to take the transmission-line solution as our starting point. This has the advantage of presenting a more physical picture of the properties and behaviour of plane waves. A combination of two plane waves is shown to describe the fields in simple rectangular waveguides. This result was used in Chapter 7 to obtain the fields radiated from rectangular horns. But it is also a simple, yet striking, example of the application of the angular plane-wave spectrum principle.

A.1 PLANE WAVES DERIVED FROM THE PARALLEL-PLATE TRANSMISSION LINE

Elementary calculations show that, if fringing effects are ignored, the capacitance C and inductance L per unit length of the parallel plate transmission line shown in Fig. A.1 are respectively

$$C = \epsilon \frac{b}{a} \tag{A.1}$$

and

$$L = \mu \frac{a}{b} \tag{A.2}$$

where ϵ and μ are the permittivity and permeability of the medium filling the space between the plates. The phase velocity v and characteristic impedance Z_c are thus

$$v = \frac{1}{\sqrt{(LC)}} = \frac{1}{\sqrt{(\epsilon\mu)}} \tag{A.3}$$

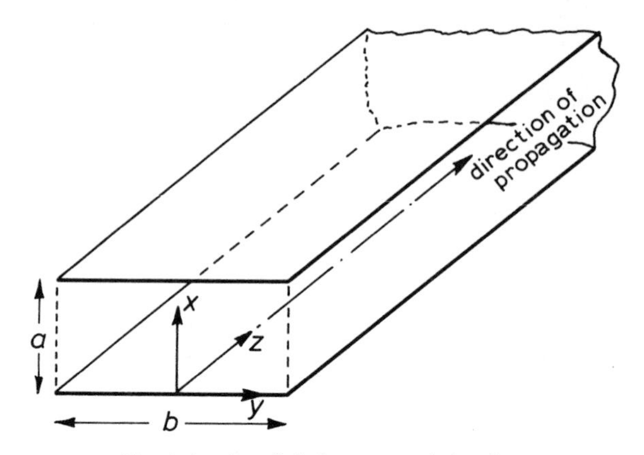

Fig. A.1 – Parallel-plate transmission line.

and

$$Z_c = \sqrt{\frac{L}{C}} = \frac{a}{b}\sqrt{\frac{\mu}{\epsilon}} \tag{A.4}$$

A wave of angular frequency ω, travelling in the z-direction, therefore has voltage and current given by

$$V = A \exp(-jkz) \tag{A.5}$$

$$I = \frac{A}{Z_c} \exp(-jkz) \tag{A.6}$$

where

$$k = \frac{\omega}{v} = \omega\sqrt{\mu\epsilon} \tag{A.7}$$

and A is a constant. The time dependent factor exp $(j\omega t)$ has been omitted, as both V and I are regarded as phasors.

The electric and magnetic fields associated with the voltage and current wave of equations (A.5) and (A.6) are easily found in the case of an ideal parallel-plate transmission line. The neglect of fringing is made more plausible by supposing that guard plates, shown by dashed lines in Fig. A.2, are included and that the voltage V is maintained between them. Then the electric field between the plates will be constant and wholly x-directed. Taking the lower plate to be at potential V with respect to the upper plate, it follows that

$$E_x = \frac{V}{a}\ , \tag{A.8}$$

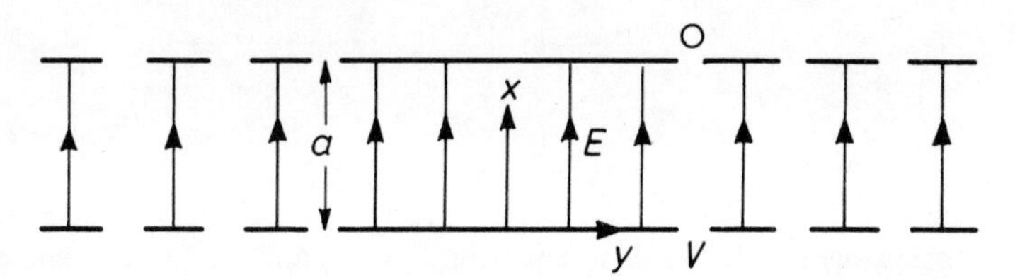

Fig. A.2 – Electric field between parallel plates. Guard plates ensure the field is uniform and x-directed.

and from equation (A.5) that

$$E_x = \frac{A}{a} \exp(-jkz) \quad . \tag{A.9}$$

The current I flows in the z-direction, and so the magnetic field, again neglecting fringing, is wholly y-directed as shown in Fig. A.3. Applying Ampère's

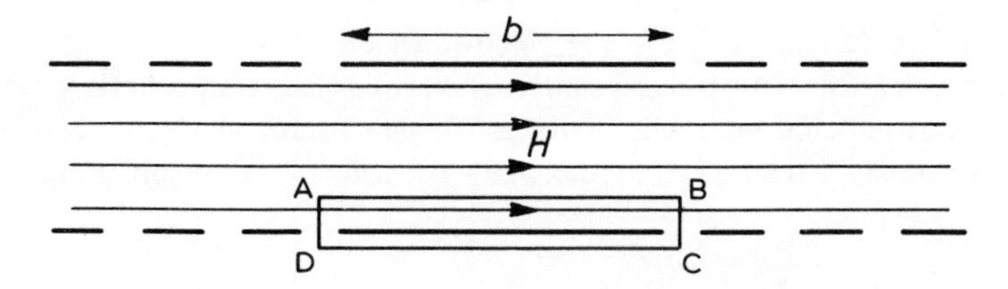

Fig. A.3 – Magnetic field between parallel plates.

law (see section B.1.4) to a rectangle such as ABCD in the figure, and noting that the magnetic field is zero outside the plates, it follows that

$$H_y b = I \tag{A.10}$$

and from equation (A.6) that

$$H_y = \frac{A}{bZ_c} \exp(-jkz) \tag{A.11}$$

A more convenient form for equations (A.9) and (A.11) is obtained by letting $(A/a) = E_o$. Then

$$E_x = E_o \exp(-jkz) \tag{A.12}$$

$$H_y = \frac{E_o}{Z_w} \exp(-jkz) \tag{A.13}$$

where

$$Z_W = \sqrt{\frac{\mu}{\epsilon}} \tag{A.14}$$

The expressions for the fields in equations (A.12) and (A.13) are remarkable in that they do not depend on the transmission-line dimensions a and b. In particular, if the separation a is increased indefinitely the electromagnetic field represented by these equations can still exist, and we conclude that it could also exist if the plates were removed altogether. We have thus found one form of electromagnetic wave which can propagate in a uniform medium. Since the fields are uniform in both amplitude and phase over any plane for which z is a constant, this particular electromagnetic wave is known as a **plane wave.** A more detailed examination reveals the physical mechanism by which the wave propagates as being one in which the time-varying magnetic field generates an electric field (Faraday's law, section B.1.3) and the time-varying electric field generates a magnetic field (Ampère's law, section B.1.4).

The above derivation of a plane electromagnetic wave is not a rigorous one, but it serves to establish a close similarity between electromagnetic waves and waves on transmission lines which will prove very useful. A rigorous derivation may be provided by starting from Maxwell's equations – the result is the same.

A.2 PROPERTIES OF PLANE WAVES

The electromagnetic wave defined by equations (A.12) and (A.13) has the following physical properties:

(a) The electric and magnetic fields are both constant over any plane at right angles to the direction of propagation.

(b) The directions of the electric field, the magnetic field and propagation are mutually orthogonal and form a right-handed set in the order stated. This means that the directions of the thumb, forefinger and middle finger of the right hand provide the directions of the electric field, magnetic field and propagation respectively.

(c) The fields change with position only in the direction of propagation: there is no amplitude change and the phase changes by $k = \omega \sqrt{(\epsilon\mu)}$ radians per unit distance.

(d) The electric and magnetic fields at any point have identical phases and their magnitudes satisfy the relation:

(magnitude of electric field) = Z_W × (magnitude of magnetic field)

where Z_W is given by equation (A.14).

The quantity Z_W has properties similar to those of the characteristic impedance of a transmission line. This quantity is associated with the plane wave and is best described as the **plane wave impedance.** In practice, this is contracted to wave impedance, but it should be remembered that the value $\sqrt{(\mu/\epsilon)}$ applies to plane waves. Other forms of electromagnetic wave exist and may have different values of wave impedance.

The reciprocal of Z_w, usually denoted Y_w, is called the **wave admittance.**

In free space, $\epsilon = \epsilon_o \approx 10^{-9}/36\pi$ and $\mu = \mu_o = 4\pi \times 10^{-7}$. The velocity is then $1/\sqrt{(\mu_o\epsilon_o)} \approx 3 \times 10^8$ m/s, the well-known value for the velocity of light The wave impedance, $\sqrt{(\mu_o/\epsilon_o)} \approx 120\pi = 377\ \Omega$. In a dielectric material of relative permittivity ϵ_r, the velocity and wave impedance are $3 \times 10^8/\sqrt{\epsilon_r}$ m/s and $377/\sqrt{\epsilon_r}\ \Omega$ respectively.

A.3 POWER FLOW IN A PLANE WAVE

If we return to the transmission line of Fig. A.1 we find that the power flow in the z-direction is $\frac{1}{2}\mathrm{Re}\ VI^*$, provided that V and I are such that $|V|$, $|I|$ are peak values. Since the electric and magnetic fields are constant over the line cross-section, the power flux, that is, power flux per unit area, will be $\frac{1}{2}\mathrm{Re}\ VI^*/(ab)$. Using equations (A.8) and (A.10), we find that the power flux is $\frac{1}{2}\mathrm{Re}\ E_xH_y{}^*$ (W/m^2). The power flux is directed in the positive z-direction and the form of this result suggests the vector form

$$\mathbf{S} = \tfrac{1}{2}\mathrm{Re}\ \mathbf{E} \times \mathbf{H}^* \qquad \text{(A.15)}$$

where **S**, usually known as **Poynting's vector**, defines the direction of the power flux and its magnitude in W/m^2. **E**, **H** are vectors which represent the directions and magnitudes of the electric and magnetic fields. Note that **E**, **H** are in general complex corresponding to voltage and current phasors. They therefore also contain phase information.

Equation (A.15) has only been derived for the special case of a plane electromagnetic wave. A fuller discussion starting from Maxwell's equations shows that **S** may be interpreted as a power flux for any form of electromagnetic wave with sinusoidal time dependence (see sections B.2.2 and B.2.5).

A.4 GENERAL EXPRESSION FOR A PLANE WAVE

The results given in section A.1 hold for the special case of a plane wave propagating in the z-direction, and with the electric and magnectic fields in the x- and y-directions respectively. We may use the physical properties listed in section A.2 to derive expressions for the general case in which the direction of propagation may have any value. We define the direction of propagation by a unit vector **u** and position in space by a position vector **r** drawn from some selected origin O. Properties (a) and (c) indicate the nature of the field dependence on position. Referring to Fig. A.4, we note that the fields are constant over the planes P_o, P_1, P_2 . . . . at right angles to **u**. For any such plane the distance from O

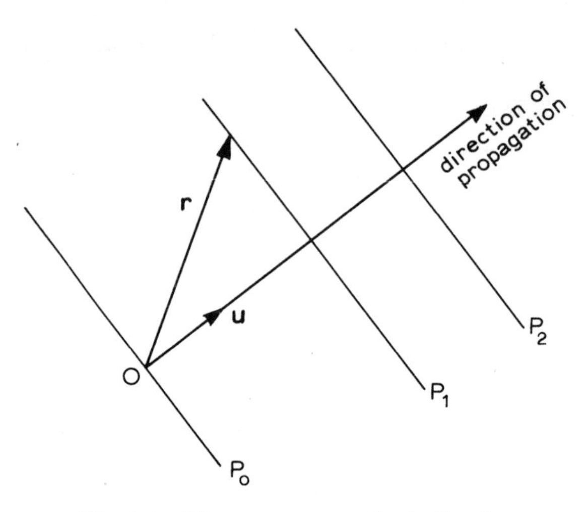

Fig. A.4 – Plane wave propagating in direction **u**.

measured in the direction **u** is simply **u.r** (remembering that **u** is a unit vector) and so the factor corresponding to exp $(-jkz)$ in section A.1 is exp $(-jk\ \mathbf{u.r})$. For this plane wave we now have

$$\mathbf{E(r)} = \mathbf{E}_o \exp(-jk\ \mathbf{u.r})$$

where the electric field is expressed as **E(r)** to indicate that it depends on position **r** and $\mathbf{E}_o$ is a constant vector, equal to the value of the electric field at any point in the plane P_o.

The vector $\mathbf{E}_o$ is not completely arbitrary since equation (b) requires that $\mathbf{E}_o$ and **u** are orthogonal, that is,

$$\mathbf{E}_o.\mathbf{u} = 0 \quad . \tag{A.16}$$

Conditions (b) and (d) may be used to obtain the corresponding expression for $H(r)$. Since **E(r)**, **H(r)** and **u** are a mutually orthogonal, right-handed set of vectors,

$$\mathbf{H(r)} = \alpha\mathbf{u} \times \mathbf{E(r)} \tag{A.17}$$

where α is a scalar. Further, since **u** and **E(r)** are orthogonal and **u** is a unit vector

$$H(r) = \alpha\, E(r) \quad .$$

But $\quad H(r) = Y_w\, E(r)$ from condition (d), giving

$$\alpha = Y_w \quad .$$

Collecting results we have as the general expressions for a plane wave propagating in the direction **u**,

$$\mathbf{E}(\mathbf{r}) = \mathbf{E}_o \exp(-jk\,\mathbf{u}.\mathbf{r}) \tag{A.18}$$

$$\mathbf{H}(\mathbf{r}) = Y_w\,\mathbf{u} \times \mathbf{E}_o \exp(-jk\,\mathbf{u}.\mathbf{r}) \tag{A.19}$$

$$\mathbf{u}.\mathbf{u} = 1 \text{ and } \mathbf{E}_o.\mathbf{u} = 0 \quad . \tag{A.20}$$

The corresponding expression for the power flux is

$$\mathbf{S} = \tfrac{1}{2}\mathrm{Re}\,\mathbf{E}(\mathbf{r}) \times \mathbf{H}^*(\mathbf{r}) = \tfrac{1}{2}\mathrm{Re}\,\mathbf{E}_o \times Y_w\,(\mathbf{u} \times \mathbf{E}_o^*)$$

$$= \tfrac{1}{2} Y_w\,(\mathbf{E}_o.\mathbf{E}_o^*)\mathbf{u} \quad . \tag{A.21}$$

In equation (A.21), the exponential factors cancel since **u** is a real vector. We have also used equation (A.16) in expanding the vector triple product as given in equation (2.13).

A.5 POLARIZATION

The electric field $\mathbf{E}_o$ is related to **u** by equation (A.16). In general, a vector has three components but in the case of $\mathbf{E}_o$ only two of these can be chosen freely, the third then being fixed by equation (A.16). This is equivalent to saying that $\mathbf{E}_o$ has two degrees of freedom. Consider again the special case of section A.1, for which $\mathbf{u} = \mathbf{u}_z$, the unit vector in the z-direction. Equation (A.16) now requires that the z-component of $\mathbf{E}_o$ should be zero. This leaves the possibility that

$$\mathbf{E}_o = \mathbf{u}_x E_x + \mathbf{u}_y E_y \tag{A.22}$$

$\mathbf{u}_x$, $\mathbf{u}_y$ being unit vectors in the x- and y-directions respectively.

The nature of the vector $\mathbf{E}_o$ defines the polarization of the wave. For the case considered in section A.1, $\mathbf{E}_o = \mathbf{u}_x E_x$, that is, the electric field has a constant direction. Such a wave is said to be **linearly polarized**, its direction of polarization being the x-axis.

A second linearly polarized wave is found by letting $\mathbf{E}_o = \mathbf{u}_y E_y$, the direction of polarization now being the y-axis.

The x- and y-linearly polarized waves are independent in the sense that both can exist with any amplitude and phase relation between them. Other forms of polarization can be obtained by combing the two linear polarizations. We consider several possibilities.

If $E_x = E_o \cos\alpha$ and $E_y = E_o \sin\alpha$,

$$\mathbf{E} = E_o\,(\mathbf{u}_x \cos\alpha + \mathbf{u}_y \sin\alpha)$$

which is linearly polarized in a direction making an angle α with the x-axis. The key point in this case is that the x- and y-components are either in phase (if $\cos\alpha$ and $\sin\alpha$ have the same signs) or in antiphase (if $\cos\alpha$ and $\sin\alpha$ have different signs). It is illustrated in Fig. A.5.

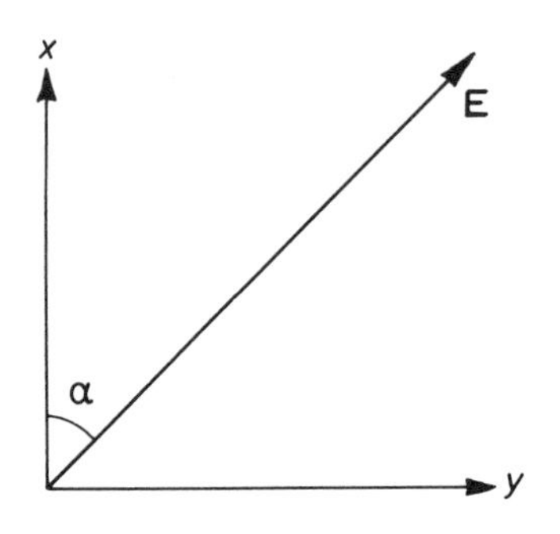

Fig. A.5 – Linearly polarized plane wave, with $E_x = E_o \cos\alpha$ and $E_y = E_o \sin\alpha$.

We now suppose that E_x, E_y are equal in magnitude but have a $\pi/2$ phase difference, that is,

$$E_x = E_o \; ; E_y = -jE_o \quad .$$

In our analysis so far we have ignored the time dependent factor $\exp(j\omega t)$ which was omitted from our starting point, the transmission line equations. In the plane $z = 0$ the full expression for the electric field, restoring the time dependence, is

$$\mathbf{E}(t) = E_o(\mathbf{u}_x - j\mathbf{u}_y)\exp(j\omega t)$$

and to obtain observable waveforms we must take the real part, that is,

$$\text{Re}\,(\mathbf{E}(t)) = |E_o|\;[\mathbf{u}_x\;\cos(\omega t + \phi) + \mathbf{u}_y\sin(\omega t + \phi)]$$

where

$$E_o = |E_o|\exp(j\phi) \quad .$$

Hence.

$$E_x = |E_o|\cos(\omega t + \phi)\; ; E_y = |E_o|\sin(\omega t + \phi)$$

and

$$E_x{}^2 + E_y{}^2 = |E_o|^2 \quad .$$

The electric field vector thus has a constant magnitude $|E_o|$ and a direction which changes with time. At time t, the vector makes an angle $(\omega t + \phi)$ with the x-axis and rotates through 2π during each period of the a.c. waveform. The vector thus traces a circle in the clockwise direction as indicated in Fig. A.6. Notice that this clockwise direction applies when the observer looks along the direction of propagation, in this case the z-axis. Such a wave is said to be a **positive**, or **clockwise, circularly polarized wave.**

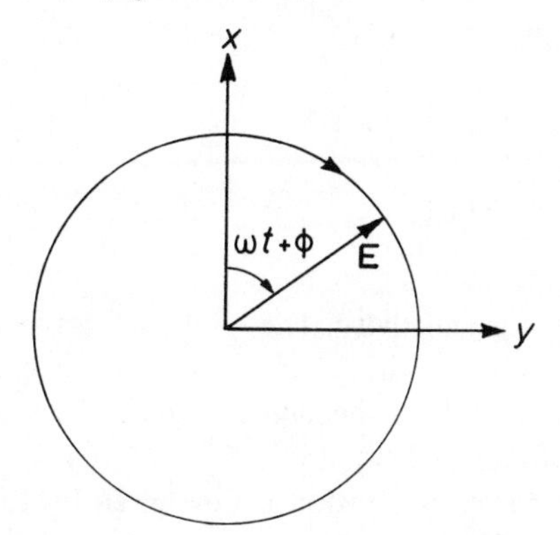

Fig. A.6 – Clockwise circular polarization: $E_x = E_o$, $E_y = -jE_o$.

If the phase difference between E_x and E_y is changed in sign, circular polarization is again obtained but with counter-clockwise rotation.

The most general case occurs when

$$\mathbf{E} = E_o\ (\mathbf{u}_x + p\mathbf{u}_y)$$

with $p = |p| \exp j\psi$ being complex. Analysis shows that the vector **E** traces an ellipse once per cycle. The major and minor axes of the ellipse have lengths $\sqrt{2}|E_o|(1 + |p|^2 + A)^{1/2}$ and $\sqrt{2}|E_o|(1 + |p|^2 - A)^{1/2}$ respectively, where $A = (1 + 2\,|p|^2 \cos 2\psi + |p|^4)^{1/2}$. The direction of the major axis makes an angle $\tan^{-1}(A - 1 + |p|^2)/(2\,|p| \cos\psi)$ with the x-axis and the sense of rotation of the vector **E** around the ellipse is clockwise when $\sin\psi < 0$ and counter-clockwise when $\sin\psi > 0$. The special case when $\sin\psi = 0$, that is, $\psi = 0$ or π, corresponds to linear polarization making an angle $\tan^{-1}(|p| \cos\psi)$ with the x-direction.

A.6 PROPAGATION IN A LOSSY MEDIUM

A simple modification to the results in section A.4 may be used when the medium in which a plane wave propagates has conductivity, σ. The parallel plate transmission model then contains a conductance G per unit length and for the dimensions defined in Fig. A.1

$$G = \sigma b/a \tag{A.23}$$

This conductance is in parallel with the capacitance, the total shunt admittance per unit length being

$$Y_s = G + j\omega C = (\sigma + j\omega\epsilon)b/a$$

which may be written as

$$Y_s = j\omega\epsilon_c b/a \tag{A.24}$$

where

$$\epsilon_c = \epsilon + \sigma/(j\omega) \tag{A.25}$$

is called the **complex permittivity**. The results for plane waves in a lossy medium follow simply by replacing the permittivity ϵ by its complex value. The phase constant and impedance both become complex. It is convenient to introduce the propagation constant, γ, equal to $\alpha + jk$ where α, k are both real. The dependence of the plane wave on position is now given by the factor $\exp(-\gamma\ \mathbf{u.r})$ or $\exp(-\alpha\ \mathbf{u.r} - jk\ \mathbf{u.r})$. The term α thus causes a progressive reduction in amplitude, that is, attenuation – as defined here it is measured in Nepers/metre. The term k is the phase change per unit distance in radians/metre as before. Replacing ϵ by ϵ_c, and using equation (A.7), we obtain:

$$\gamma = \alpha + jk = j\omega(\epsilon_c\mu)^{1/2}$$

$$= [j\omega\mu(\sigma + j\omega\epsilon)]^{1/2} \quad . \tag{A.26}$$

The corresponding expression for Z_w is

$$Z_w = [j\omega\mu/(\sigma + j\omega\epsilon)]^{1/2} \quad . \tag{A.27}$$

Equations (A.26) and (A.27) may be expressed in terms of real and imaginary parts, but the results are rather complicated. In practice, we can usually use simpler expressions appropriate to one or other of the two special cases which occur most frequently.

(a) *Metals*
For the frequencies in which we are interested, that is, for $\omega < 10^{12}$, the conductivity σ of a metal is much greater than $\omega\epsilon$. Metals typically have values of σ of 10^6 S/m and above and values of ϵ approximately equal to ϵ_0, that is, $10^{-9}/(36\pi)$

F/m. Then $\sigma/(\omega\epsilon_o) > 36\pi \times 10^3$, or 10^5 approximately. It is then legitimate to neglect the term $\omega\epsilon$ in comparison with σ in equations (A.26) and (A.27) giving:

$$\gamma = \alpha + jk = (j\omega\mu\sigma)^{1/2}$$

$$= (\omega\mu\sigma/2)^{1/2}\ (1 + j)$$

that is, $\alpha = k = (\omega\mu\sigma/2)^{1/2}$, (A.28)

and

$$Z_w = (j\omega\mu/\sigma)^{1/2} = [\omega\mu/(2\sigma)]^{1/2}(1 + j) \quad . \tag{A.29}$$

In each equation we have used the result

$$j^{1/2} = (1 + j)/\sqrt{2} \quad ,$$

which is readily checked by squaring both sides.

The amplitudes of the plane wave travelling in the metal decays proportionally to exp $(-\alpha\,\mathbf{u.r})$. The distance, δ, in which the amplitude is reduced by e is called the skin depth and it is seen that

$$\delta = \frac{1}{\alpha} = \left(\frac{2}{\omega\mu\sigma}\right)^{1/2} \tag{A.30}$$

The skin depth is very small: for example, in copper $\sigma \approx 6 \times 10^7$ S/m, and so at a frequency of 10^{10} Hz (free space wavelength, 3cm) $\delta = 0.65\ \mu$m.

An expression for Z_w which is sometimes convenient is obtained from equations (A.29) and (A.30):

$$Z_w = (1 + j)/(\sigma\delta) \quad . \tag{A.31}$$

(b) *Dielectrics*

Most dielectrics have small conductivities so that $\sigma/(\omega\epsilon)$ is much less than unity. We usually work with this ratio which is called the **loss tangent**, denoted by tan δ. (The use of δ for both skin depth in metals and the loss angle in dielectrics is regretted – there are not enough letters to cover everything). Typical values for tan δ at microwave frequencies are in the range 10^{-3} to 10^{-1}.

From equation (A.26),

$$\gamma = \alpha + jk = [-\omega^2\mu\epsilon\,(1 - j\tan\delta)]^{1/2}$$

$$\approx j\omega(\epsilon\mu)^{1/2}\ (1 - \tfrac{1}{2}\,j\tan\delta) \tag{A.32}$$

giving

$$k = \omega(\epsilon\mu)^{1/2} \text{ and } \alpha = \tfrac{1}{2}\omega(\epsilon\mu)^{1/2} \tan\delta = \tfrac{1}{2} k \tan\delta \quad .$$

These results are accurate to within 1% if $\tan\delta < 0.1$.
Similarly,

$$\begin{aligned} Z_{\mathrm{w}} &= [j\omega\mu/(j\omega\epsilon + \sigma)]^{1/2} \\ &= (\mu/\epsilon)^{1/2}/(1 - j\tan\delta)^{1/2} \qquad \text{(A.33)} \\ &= (\mu/\epsilon)^{1/2}(1 - \tfrac{1}{2} j \tan\delta) \quad . \end{aligned}$$

A.7 REFLECTION: NORMAL INCIDENCE

A number of problems of practical interest, for example reflection of waves by the land or sea surface, design of radomes, etc., involve reflection at a boundary between two media of different electrical constants. The simplest case in which a plane wave is incident normally on to a plane interface is shown in Fig. A.7 (a). This case may be reduced to an equivalent transmission line problem by reversing the arguments in section A.1. The incident wave may be represented by a corresponding wave on a transmission line with phase constant (or propagation constant if the medium is lossy) and characteristic impedance equal to the phase constant and wave impedance for the plane wave. For any value of z, the

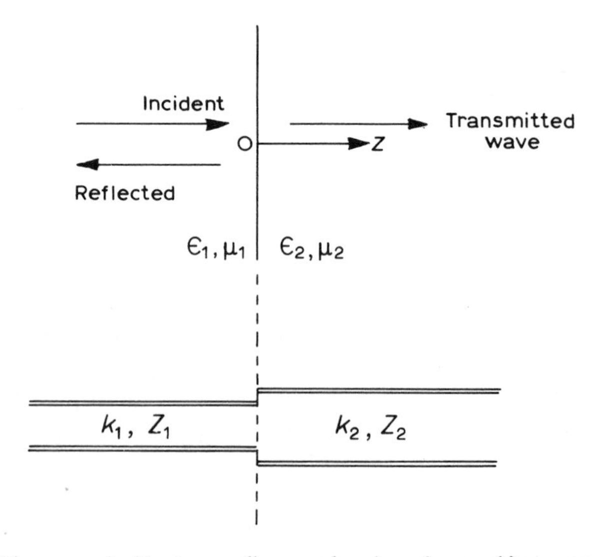

Fig. A.7 – Plane wave incident normally on a plane boundary; and its transmission-line representation.

values of the electric and magnetic fields for the incident plane wave may then be taken as numerically equal to the voltage and current of the incident wave for the same value of z on the line. Similarly, a plane wave transmitted into the second medium may be represented by a wave on a second transmission line.

The question then arises as to how the two lines should be joined. This is answered by considering the boundary conditions which apply on the interface $z = 0$ in Fig. A.7(a). These boundary conditions, derived from Ampère's law and Faraday's law (see section B.2.3) are:

(i) the components of the electric field tangential to the interface are continuous

and

(ii) the components of the magnetic field tangential to the interface are continuous.

The equality of the electric and magnetic fields in the media to the corresponding voltages and currents then leads to the conclusion that the lines must be connected to retain continuity of voltages and current at $z = 0$. The direct connection of the lines as in Fig. A.7(b) ensures this continuity.

In the transmission line circuit, a reflected wave travelling in the negative z-direction is established on line 1, the reflection coefficient at $z = 0$ being

$$\rho = \frac{Z_2 - Z_1}{Z_2 + Z_1} \tag{A.34}$$

A corresponding reflected plane wave appears in medium 1. If either medium is lossy, one of the impedances and hence ρ will be complex. The argument of ρ gives the phase difference between the incident and reflected waves at $z = 0$.

The complex amplitude of the wave transmitted into medium 2 is τE at $z = 0$ where E is the incident field at $z = 0$ and

$$\tau = \frac{2Z_2}{Z_1 + Z_2} \tag{A.35}$$

Note that if the second medium is a metal the wave will penetrate to a depth of the skin-depth. When the metal is perfectly conducting and σ becomes infinite the current density in the metal also becomes infinite. But the corresponding skin-depth then becomes zero. Hence the current density in the metal becomes a surface current density of finite magnitude, given by simple transmission-line theory as twice the magnitude of the magnetic field of the normally incident plane wave.

A.8 REFLECTION AND REFRACTION

A more general case of a plane wave incident on the interface between two media is shown in Fig. A.8(a), the direction of propagation of the incident wave

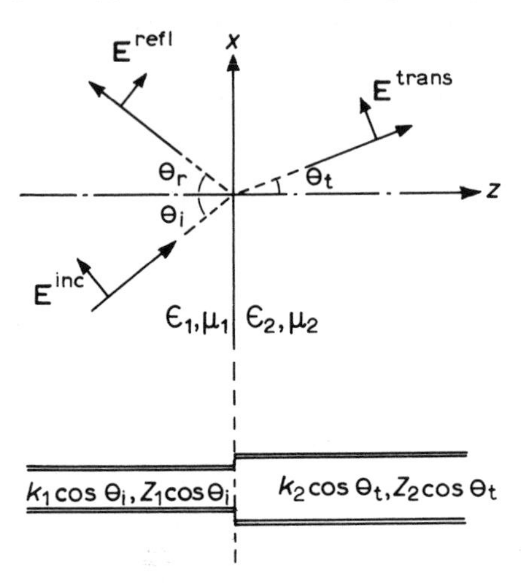

Fig. A.8 – Plane wave incident obliquely on a plane boundary; and its transmission-line representation.

now making an angle θ_i with the normal to the interface: θ_i is called the **angle of incidence.** The electric field of the incident wave may be written from equation (A.18) as:

$$\mathbf{E}^{\text{inc}} = \mathbf{E}_o \exp\{-jk_1(x \sin\theta_i + z\cos\theta_i)\} \tag{A.36}$$

where $\mathbf{E}_o$ must be chosen to satisfy equation (A.20). The appearance of two coordinates x,z in the expression for $\mathbf{E}^{\text{inc}}$ leads to some complication, but we will find that an equivalent transmission-line circuit may still be found. The key step is to note that the continuity of tangential electric and magnetic fields must hold at every point on the interface $z = 0$, that is, for all values of x. This implies that the dependence on x of both the reflected and transmitted plane waves must be the same as that of the incident wave. In other words the factor $\exp(-jk_1 x \sin\theta_i)$ must be common to all three plane waves.

Consider a reflected wave making an angle of reflection θ_r with the normal. Its electric field will be

$$\mathbf{E}^{\text{refl}} = \mathbf{E}_1 \exp\{-jk_1(x\sin\theta_r - z\cos\theta_r)\} \quad . \tag{A.37}$$

Note the sign of the z-dependent term, with the wave travelling towards the negative z-direction. From the argument above we require

$$\exp(-jkx\sin\theta_i) = \exp(-jkx\sin\theta_r)$$

that is,

$$\sin \theta_i = \sin \theta_r$$

leading to the well-known optical result for reflection:

$$\text{angle of incidence, } \theta_i = \text{angle of reflection, } \theta_r \quad . \tag{A.38}$$

The reflected wave becomes

$$\mathbf{E}^{\text{refl}} = \mathbf{E}_1 \exp \{-jk_1 (x \sin \theta_i - z \cos \theta_i)\} \quad . \tag{A.39}$$

Since the x-dependent factor is common to equations (A.36) and (A.39), we may concentrate on the z-dependence and choose a transmission line to represent this. The phase constant for this line must therefore be $k_1 \cos \theta_i$ to give z-dependent factors of $\exp(-jk_1 z \cos \theta_i)$ and $\exp(+jk_1 z \cos \theta_i)$ for the incident and reflected waves respectively.

We now consider the transmitted wave and suppose its direction of propagation makes an angle θ_t with the z-axis: θ_t is called the **angle of transmission** (or **refraction**). The electric field of this wave is

$$\mathbf{E}^{\text{trans}} = \mathbf{E}_2 \exp \{-jk_2 (x \sin \theta_t + z \cos \theta_t)\} \tag{A.40}$$

and again we require the x-dependence to be $\exp(-jk_1 x \sin \theta_i)$. We therefore have

$$k_1 \sin \theta_i = k_2 \sin \theta_t \tag{A.41}$$

which may be recognised as Snell's law. It can be made identical to Snell's law if we define the refractive index n of a medium as the ratio of its phase constant k to the phase constant k_o for a plane wave of identical frequency propagating in free space. Then,

$$n = \frac{k}{k_o} = \sqrt{\frac{\epsilon\mu}{\epsilon_o \mu_o}} \quad , \tag{A.42}$$

and equation (A.41) becomes

$$n_1 \sin \theta_i = n_2 \sin \theta_t \quad . \tag{A.43}$$

For lossy media, we use γ in place of k and find that the refractive index becomes complex, and is usually also frequency-dependent.

The transmission line to represent the z-dependence of the transmitted wave must have a phase constant, $k_2 \cos \theta_t$.

The results so far are general in that no restriction has been placed on the polarization of the incident wave. However, to proceed further leading to a transmission line circuit from which we can calculate $\mathbf{E}_1, \mathbf{E}_2$, we require the values of the impedances. We obtain these for two special cases of the polarization, noting that the general case may always be tackled by expressing the incident wave as a combination of these two cases. The choice of the two special cases is governed by the need to satisfy the boundary conditions which apply to the field components tangential to the interface.

E-plane case

In the first case, we take the polarization of the incident wave to be such that its electric field lies in the plane of Fig. A.8(a) – hence the name E-plane case. This electric field has both x- and z-components, and since it is normal to the direction of propagation we find

$$E_x^{\text{inc}} = E_o \cos \theta_i \quad . \tag{A.44}$$

Similarly for the electric field of the reflected and transmitted waves,

$$E_x^{\text{refl}} = E_1 \cos \theta_i \tag{A.45}$$

$$E_x^{\text{trans}} = E_2 \cos \theta_t \quad . \tag{A.46}$$

The magnetic fields of all three plane waves have y-components only, that is,

$$H_y^{\text{inc}} = H_o = E_o/Z_1 \tag{A.47}$$

$$H_y^{\text{refl}} = -H_1 = -E_1/Z_1 \tag{A.48}$$

$$H_y^{\text{trans}} = H_2 = E_2/Z_2 \quad . \tag{A.49}$$

The sign reversal for the reflected wave arises since this wave is travelling towards the negative z-direction (as may be seen by using the right hand rule).

At $z = 0$, the tangential components are E_x, H_y which must be continuous: direct connection of the transmission lines gives continuity of voltage and current. We conclude that equivalence of the circuit in Fig. A.8(b) to the situation in Fig. A.8(a) requires that voltage and current equal E_x and H_y respectively. The impedance of the line 1, Z_1', is thus given by the ratio E_x/H_y for the incident plane wave, giving

$$Z_1' = Z_1 \cos \theta_i \quad , \tag{A.50}$$

and similarly for the second line

$$Z_2' = Z_2 \cos \theta_t \quad . \tag{A.51}$$

The reflection coefficient is therefore

$$\rho = \frac{Z_2' - Z_1'}{Z_2' + Z_1'} = \frac{Z_2 \cos \theta_t - Z_1 \cos \theta_i}{Z_2 \cos \theta_t + Z_1 \cos \theta_i} \quad , \tag{A.52}$$

with θ_t being obtained from equation (A.41).

We have not considered the field components normal to the interface, that is, E_z and H_z. This is the case in all electromagnetic wave boundary problems – look after the tangential components, and the normal components will look after themselves.

The reflection coefficient, ρ, in equation (A.52) refers to the x-components of the electric fields of the incident and reflected plane waves, that is,

$$E_{1x} = \rho E_{ox} \quad . \tag{A.53}$$

We also require the component E_{1z} and obtain this by noting that

$$\frac{E_{1z}}{E_{1x}} = -\tan \theta_i \quad , \tag{A.54}$$

since $\mathbf{E}_1$ is normal to the direction in which the reflected wave propagates. It is easily shown that

$$E_{1z} = -\rho E_{oz} \quad . \tag{A.55}$$

The electric field for the transmitted wave is found by a similar argument.

$$E_{2x} = \tau E_{ox} \text{ where } \tau = 1 + \rho \tag{A.56}$$

and

$$E_{2z} = E_{2x} \tan \theta_t \quad . \tag{A.57}$$

H-plane case

In the second case, the magnetic field lies in the plane of the diagram having x- and z-components with only a y-component of electric field. The argument follows the same lines as previously with impedance in the lines made equal to

$-E_y/H_x$, the minus sign arising because of the right hand relation. The line impedances are therefore

$$Z_1'' = Z_1 \sec \theta_i \; ; \; Z_2'' = Z_2 \sec \theta_t \quad . \tag{A.58}$$

A.9 TOTAL INTERNAL REFLECTION

The analysis in the preceding sections has proceeded on the assumption that the angles θ_i, θ_r, θ_t are all real. Examination of the argument shows, however, that it remains valid provided the propagation vector **u** introduced in equation (A.18) satisfies $\mathbf{u.u} = 1$. For the transmitted wave,

$$\mathbf{u} = \sin \theta_t \, \mathbf{u}_x + \cos \theta_t \, \mathbf{u}_z$$

and so we need only that

$$\sin^2 \theta_t + \cos^2 \theta_t = 1.$$

Now, suppose medium 1 is a dielectric with electrical constants ϵ_1, μ_o, where $\epsilon_1 > \epsilon_o$, and medium 2 is free space. Snell's law then reduces to

$$(\epsilon_r)^{1/2} \sin \theta_i = \sin \theta_t$$

where $\epsilon_r = \epsilon_1/\epsilon_o$ and $\sin \theta_1 > 1$ if $\sin \theta_i > (\epsilon_r)^{-1/2}$. The angle of refraction is no longer real, so what happens? Consider $\cos \theta_t$:

$$\cos^2 \theta_t = 1 - \sin^2 \theta_t < 0$$

so that $\cos \theta_t$ becomes imaginary, say $\pm j\chi$, with χ real and positive.

The transmitted plane wave is now:

$$\mathbf{E}^{\text{trans}} = \mathbf{E}_2 \exp \{-jk_o \, x \sin \theta_t \pm \chi k_o z\}$$

and thus shows either an exponential growth or decay as we move away from the interface at $z = 0$. Since this wave originates from the incident plane wave in medium 1, it is physically very unlikely that exponential growth will occur as z increases. The physically possible solution must therefore be proportional to $\exp(-\chi k_o z)$ which vanishes as z increases. Such a wave is said to be **evanescent.** The required value of $\cos \theta_t$ is $-j\chi$.

The line impedance corresponding to the evanescent wave is $-j\chi Z_o$ or jZ_o/χ, for the E-plane and H-plane cases respectively. In both cases this impedance is imaginary and we find $|\rho| = 1$, that is, the incident wave is totally reflected at the interface. This result, total internal reflection, is well known in optics and has been used at millimetric wavelengths to provide useful devices such as attenuators. Its most important application may well be in optical fibres.

A.10 WAVEGUIDES

In this section, we will explore the behaviour of an electromagnetic wave propagating in a perfectly conducting cylinder of rectangular cross-section – a rectangular waveguide. We begin by examining the behaviour of plane waves reflected by each of the two perfectly conducting planes shown in Fig. A.9(a). Suppose two such plane waves exist, propagating in directions making angles of $\pm\,\theta$ with the y-axis. The x-axis is assumed to be normally inwards to the plane of the diagram and the y-axis as shown. This choice of axes has been made to retain for the final solution a correspondence between the directions of the electric and magnetic fields of propagation and the x-, y- and z-axes similar (but not identical as will be seen later) to that for the plane wave considered in section A.1.

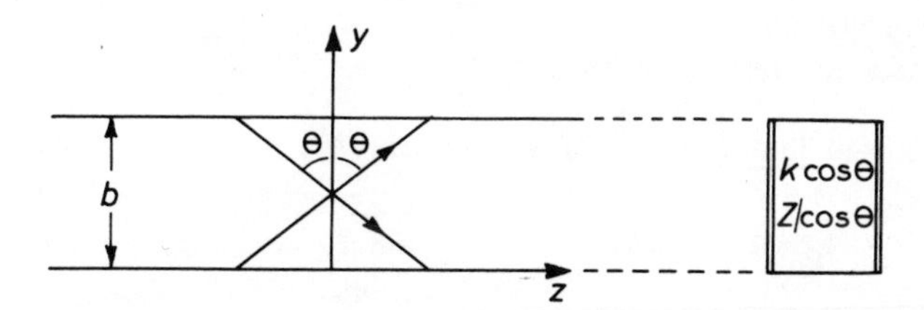

Fig. A.9 – Two plane waves reflected by two perfectly conducting planes; and its transmission-line representation.

The medium between the plates has constants ϵ,μ – in practice we will usually be considering free space. The direction of the electric field for either plane wave is the x-direction. We are thus considering an H-plane situation. For this configuration, the z-dependence of both plane waves is the same, $\exp(-jkz\sin\theta)$, and we may use a transmission line circuit to represent the behaviour in so far as dependence on y is concerned. The circuit is shown in Fig. A.9(b): the phase constant is $k\cos\theta$ and the impedance, obtained from the ratio $-E_x/H_z$ for the wave travelling towards the positive y-direction, is $Z\sec\theta$. The voltage and current on the transmission line correspond to E_x and $-H_z$ respectively ($-H_z$ from right hand rule). The terminations of the line at $y = 0$ and $y = b$ are both short circuits, since the impedance of the metal walls is zero when σ is infinite.

The circuit in Fig. A.9(b) has no sources, and voltages and currents may exist only if the circuit is resonant. The voltage has the general form

$$V = A\exp(-jky\cos\theta) + B\exp(jky\cos\theta) \tag{A.58}$$

and the resonance condition is that $V = 0$ when $y = 0$ or b. Then

$$A + B = 0 \text{ for } y = 0$$

and

$$A \exp(-jkb \cos\theta) + B \exp(jkb \cos\theta) = 0,$$

so that

$$B = -A \tag{A.59}$$

and

$$\sin(kb \cos\theta) = 0 \quad . \tag{A.60}$$

In equation (A.60), b and k are fixed, if the frequency is fixed, and so we find that the assumed combination of plane waves exists only for the values of θ satisfying

$$kb \cos\theta = m\pi \quad , \quad m = 0, 1, 2, \ldots \tag{A.61}$$

The value of $m = 0$ does not lead to a physically possible solution since the corresponding line impedance $Z \sec\theta$ is infinite. Each other value of m gives a 'mode' of propagation.

From the above results, we find

$$V = A \,[\exp(-jm\pi y/b) - \exp(jm\pi y/b)]$$

$$= A_o \sin(m\pi y/b) \tag{A.62}$$

where

$$A_o = -2jA \tag{A.63}$$

and

$$I = A\,[\exp(-jm\pi y/b) + \exp(jm\pi y/b)]\,/(Z \sec\theta)$$

$$= \frac{jA_o m\pi}{Zkb} \cos(m\pi y/b) \quad . \tag{A.64}$$

Recalling

(a) that V,I correspond to E_x, $-H_z$ respectively

and

(b) the factor $\exp(-jkz\sin\theta)$, we may use equations (A.62) and (A.64) to obtain expressions for the fields within the plates. We first use equation (A.61) to obtain an expression for $k \sin\theta$. We denote $jk \sin\theta$ by γ_m. Then,

$$\gamma_m = j\beta_m = \left[k^2 - \left(\frac{m\pi}{b}\right)^2\right]^{1/2} \text{ if } \frac{m\pi}{b} < k \tag{A.65}$$

$$\gamma_m = \alpha_m = \left[\left(\frac{m\pi}{b}\right)^2 - k^2\right]^{1/2} \text{ if } \frac{m\pi}{b} > k \ , \tag{A.66}$$

and

$$E_x = A_o \sin(m\pi y/b) \exp(-\gamma_m z) \quad . \tag{A.67}$$

$$H_z = \frac{jm\pi A_o}{kbZ} \cos(m\pi y/b) \exp(-\gamma_m z) \quad . \tag{A.68}$$

There is in addition a y-component of magnetic field. Reference to Fig. A.9(a), and use of the right hand rule shows that $H_y = E_x \sin\theta/Z$ for each of the two plane waves and therefore this relation still applies when the two are combined. Further, $\sin\theta = \gamma m/jk$, so

$$H_y = \frac{\gamma_m A_o}{jkZ} \sin(m\pi y/b) \exp(-\gamma_m z) \quad . \tag{A.69}$$

Equations (A.65) to (A.69) specify completely the mth mode.

These results are also applicable if the parallel-plate structure is converted into a rectangular waveguide by the addition of conducting planes at $x = 0$ and $x = a$. These planes are at right angles to the electric field and so do not disturb the fields.

We next consider the physical properties of the modes.

(i) The mode propagates in the z-direction with a phase change β_m provided $kb > m\pi$. Since $k = \omega(\epsilon\mu)^{1/2}$, there is a frequency $\omega_m = m\pi/b(\epsilon\mu)^{1/2}$ such that propagation without attenuation occurs only when ω exceeds ω_m. The term **cut-off frequency** is applied to ω_m. When $\omega < \omega_m$, the wave is evanescent, that is, it is attenuated. Equations (A.65) and (A.66) are often more conveniently expressed in frequency terms:

$$\beta_m = (\epsilon\mu)^{1/2} (\omega^2 - \omega_m{}^2)^{1/2} \text{ for } \omega > \omega_m \tag{A.70}$$

$$\alpha_m = (\epsilon\mu)^{1/2} (\omega_m{}^2 - \omega^2)^{1/2} \text{ for } \omega < \omega_m \quad . \tag{A.71}$$

We note that the cut-off frequency is directly proportional to the mode number. The behaviour of these modes as a function of frequency is best shown graphically as in Fig. A.10. The frequency range $\omega_1 < \omega < \omega_2$ is of particular interest since only one mode, that for $m = 1$, propagates in this range. This mode is often referred to as **dominant**. Waveguides are usually operated so that the dominant mode only is propagating.

(ii) Each mode has three field components E_x, H_y, H_z. The first two are normal or transverse to the direction of propagation but the third is in the propagation direction. We thus have a wave that is different from a simple plane wave, even though it is obtained by combining plane waves. The following nomenclature is used. The plane wave has fields which are entirely transverse to the propagation direction and is described as a **transverse electromagnetic** (TEM) wave. The modes we have now found have the electric field transverse to the propagation direction and are called **transverse electric** (TE) waves. It is convenient to label them so that the different values of m can be recognised. The notation TE_{0m} implies that the electric field has no variation in the x-direction and m half-cycles of variation in the y-direction. We need a double suffix since we could have a mode with field components E_y, H_x, H_z, for example, TE_{n0}.

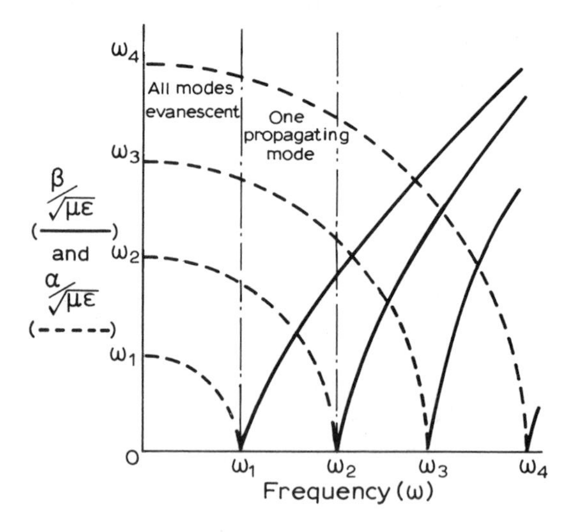

Fig. A.10 – Modes in a rectangular guide.

The TE_{10} mode would propagate if $\omega > \pi/a\ (\epsilon\mu)^{\frac{1}{2}}$. To ensure single mode propagation in the rectangular guide we therefore choose the dimension a so that the cut-off frequency of the TE_{10} mode is not less than that of the TE_{02}. This requires $a < \frac{1}{2}b$.

(iii) The transverse components of electric and magnetic fields satisfy

$$E_x = Z_g H_y$$

where

$$Z_g = \frac{jkZ}{\gamma_m} \quad , \tag{A.72}$$

a result which applies to any TE mode in a waveguide provided the guide is completely filled by a single material.

When the mode propagates, that is, $\gamma_m = j\beta_m$, then Z_g is real (as in an ordinary transmission line). If β_m, k are expressed as $2\pi/\lambda_g$, $2\pi/\lambda_o$, where λ_g and λ_o are respectively the wavelengths of the mode and of a plane wave in the medium filling the guide, then

$$Z_g = \frac{\lambda_g Z}{\lambda_o} \tag{A.73}$$

For an evanescent mode, $\gamma_m = \alpha_m$ and

$$Z_g = \frac{jkZ}{\alpha} = \frac{j\omega\mu}{\alpha}, \tag{A.74}$$

a positive, or inductive, reactance.

A.11 CONCLUDING REMARK

This appendix has been designed to provide an introduction to electromagnetic waves without the mathematical complexities introduced by Maxwell's equations. Any valid field must satisfy these equations, and in the situations with which we are concerned it is sufficient (see section B.2.5) to use the two equations

$$\nabla \times \mathbf{H} = j\omega\epsilon\mathbf{E} \tag{A.75}$$

$$\nabla \times \mathbf{E} = -j\omega\mu\mathbf{H} \tag{A.76}$$

together with the boundary conditions. The validity of all the solutions obtained in this appendix may be checked by showing that they do indeed satisfy these equations, which is an easier task than deriving the solutions directly from the equations.

Appendix B

Summary of Maxwell's equations and some of their consequences

The following summary is intended as a reminder, and as a handy list for reference, of the main results of electromagnetic theory. The reader will have his own favourite texts† which give the development of the theory in detail. The exercises in this appendix are used as an artificial means of filling in some of this missing detail.

B. 1 MAXWELL'S EQUATIONS

The equations set down by James Clerk-Maxwell in 1873, and generally known as **Maxwell's equations**, govern the phenomena of electricity and magnetism on a macroscopic scale (that is, scale sizes which are very large compared to atomic dimensions). There are traditionally four equations, which can be linked to, and considered to be mathematical generalizations of, experimental observations which have acquired the status of physical laws. The equations and their attendant definitions will be introduced in turn by way of these physical laws. It will be understood that the validity of Maxwell's equations is attested by countless experiments and by the precision so often attained in electrical technology.

B.1.1 Electric Field. Divergence of Electric Flux Density

An electric field is a force experienced by an electric charge. In the case of static fields (no time variation) this force is due to the presence of other electric charges.

The experimental law governing the forces between electric charges is **Coulomb's law**. The magnitude of the force which would be experienced by unit charge (1 coulomb) is the electric field strength E(volts per metre), which is the magnitude of the vector electric field **E**.

Coulomb's law (see Exercise B.1.1(a)) states that the force between charges,

†Ours include Jordan and Balmain (1968), Plonsey and Collin (1961), Ramo, Whinnery and Van Duzer (1965) and Stratton (1941).

and hence electric field, depends inversely on an electrical property of the medium known as the permittivity ϵ (farads per metre). Thus the electric flux density **D** (coulombs per square metre) defined as

$$\mathbf{D} = \epsilon \mathbf{E} \tag{B.1.1}$$

is independent of the medium in which the charges exist, and its greater generality is often useful. In a uniform and isotropic medium the permittivity is a scalar constant, which may be expressed as

$$\epsilon = \epsilon_r \epsilon_o \tag{B.1.2}$$

where ϵ_o is the permittivity of free space, which is 8.854×10^{-12} farads per metre, and ϵ_r is the relative permittivity. In a nonuniform but istropic medium ϵ is a scalar function of position. In an anisotropic medium the permittivity is a tensor, and if the medium is also nonuniform the tensor will change with position.

Coulomb's law and the definition of **D** lead to **Gauss's law**, which states that the total electric flux passing out through a closed surface is equal to the total electric charge enclosed. Stated in mathematical terms,

$$\oint_A \mathbf{D.n}\, da = \int_V \rho\, dV \tag{B.1.3}$$

where $\mathbf{n}da$ is a vector element of the area A which encloses the volume V, and is by convention directed outward from the volume. The quantity ρ is the distribution of charge density.

A more useful form of equation (B.1.3) is obtained by applying it to a vanishingly small volume dV. If the **divergence** of the vector **D** is defined as the limit as dV goes to zero of the left hand side of equation (B.1.3) divided by dV, then

$$\text{div}\, \mathbf{D} \equiv \nabla.\mathbf{D} = \rho \quad . \tag{B.1.4}$$

This will be referred to as the **first of Maxwell's equations.** Used in conjunction with equation (B.1.1) it relates the electric field to the electric charge in a simple and convenient manner. (The vector differential operator del is defined in rectangular coordinates as

$$\nabla \equiv \mathbf{u}_x \frac{\partial}{\partial x} + \mathbf{u}_y \frac{\partial}{\partial y} + \mathbf{u}_z \frac{\partial}{\partial z} \quad ,$$

where $\mathbf{u}_x$, $\mathbf{u}_y$ and $\mathbf{u}_z$ are unit vectors in the x,y and z directions.)

Exercises B.1.1

(a) Given Coulomb's law, that the force **f** experienced by a charge q_2 a distance **r** away from a charge q_1 is

$$\mathbf{f} = \frac{q_1 q_2}{4\pi\epsilon r^2}\,\mathbf{u}_r$$

where ϵ is the permittivity of the medium and the unit vector $\mathbf{u}_r = \mathbf{r}/r$, prove Gauss's law.

(b) Show that the electric field inside a hollow charged sphere is zero.

(c) By considering an element of volume in rectangular coordinates, show that the divergence of the vector D,

$$\text{div}\,\mathbf{D} = \frac{\partial D_x}{\partial x} + \frac{\partial D_y}{\partial y} + \frac{\partial D_z}{\partial z}$$

and hence that

$$\nabla.\mathbf{D} = \rho$$

(d) If the permittivity of a medium is a scalar function of position, find the relationship between charge distribution and the distribution of electric field.

B.1.2 Magnetic Field. Divergence of Magnetic Flux Density

A magnetic field is a force experienced by a current element. In the case of static fields this force is due to the presence of other current elements. Basically, an electric current is a flow of charge, and so the magnetic field due to certain magnetic materials is not excluded from this definition since it is the motion of electrons within the atoms of such materials that gives rise to their magnetic field.

The experimental law governing the forces between current elements is generally ascribed to Biot and Savart. The magnetic flux density **B** (webers per square metre) may be thought of as having a magnitude which is equal to the force experienced by a unit current element placed in that field with an orientation such that the force is a maximum. Although the concept of a magnetic field can be stated in this way, which emphasizes the analogy with electric field, the directions associated with the forces and hence with the fields are not as simple as in the case of the electric field. Consider the example of the electric field due to a positive point charge. The field is everywhere directed radially away from the charge, and the idea that electric flux is emanating from a source, which is the charge itself, is a natural one and is embodied in the mathematical divergence relation (B.1.4). Now in the case of the magnetic field due to a current element (see exercise B.1.2(a)) the field is always directed perpendicular to the plane containing the element and the radius vector from the current element to the field point. Therefore magnetic field lines are closed on themselves, and from this purely intuitive point of view it would seem reasonable to conclude that the outward flux of magnetic field through the surface of any completely enclosed volume is always zero, or in mathematical terms

$$\oint_A \mathbf{B}.\mathbf{n}\, da = 0 \quad . \tag{B.1.5}$$

Using the definition of divergence in section B.1.1, this integral relation may be expressed in differential form as

$$\text{div}\, \mathbf{B} \equiv \nabla.\mathbf{B} = 0 \quad . \tag{B.1.6}$$

This will be referred to as the **second of Maxwell's equations.**

It will be noticed from the definition of magnetic flux density given in exercise B.1.2(a) that **B** depends directly on a magnetic property of the medium known as the permeability μ (henries per metre). It is therefore possible to define a quantity **H**, the vector magnetic field (amperes per metre), which is related to the magnetic flux density by the equation

$$\mathbf{B} = \mu\mathbf{H} \quad . \tag{B.1.7}$$

The permeability μ may be a scalar or a tensor; its value in free space is $\mu_o = 4\pi \times 10^{-7}$ henries/metre and it is often convenient to express the permeability of a medium in terms of that of free space by means of the relation

$$\mu = \mu_r\, \mu_o \tag{B.1.8}$$

where μ_r is the relative permeability of the medium.

Exercises B.1.2

(a) The elemental force **df** experienced by a current element I**dl** a distance **r** away from another current element I'**dl**$'$, based on the experiments of Biot and Savart and Ampère, is conveniently expressed by the following two vector relations

$$\mathbf{df} = I\mathbf{dl} \times \mathbf{B}$$

$$\mathbf{B} = \mu \frac{I'\mathbf{dl}' \times \mathbf{u}_r}{4\pi r^2}$$

where μ is the permeability of the medium. (Note that the intermediate quantity, the magnetic field **B**, is considered to be due to the (source) current element $I'\mathbf{dl}'$ and that this field interacts with the (test) current element $I\ \mathbf{dl}$ to produce the force $\mathbf{df}$).

Find the magnitude of the force exerted by $\mathbf{I}'\mathbf{dl}'$ on $\mathbf{I}\ \mathbf{dl}$ if the former is inclined at an angle ϕ to $\mathbf{r}$, and the latter is inclined at an angle θ to the plane containing $I'\ \mathbf{dl}'$ and $\mathbf{r}$. Indicate the direction of the force in a sketch.

(b) Calculate the force per unit length exerted on each other by two infinitely long parallel conductors situated in air a distance d apart and carrying currents I_1 and I_2.

(c) Find the magnetic field $\mathbf{H}$ at a distance r from an infinitely long conductor carrying a current I.

B.1.3 Time Variation of Magnetic Field

The first of Maxwell's equations describes the static electric field in terms of its sources, and the second describes the static magnetic field. Each equation is independent of the other. However, when they vary in time the electric field and magnetic field *are* related. The third and fourth of Maxwell's equations describe this inter-relationship.

Faraday discovered that the electromotive force around a circuit is proportional to the rate of change in time of the magnetic flux linking that circuit. In mathematical terms this may be written as the line integral

$$\oint_C \mathbf{E}.\mathbf{dl} = -\frac{\partial}{\partial t}\int_A \mathbf{B}.\mathbf{n}\ \mathrm{d}a \tag{B.1.9}$$

where $\mathbf{dl}$ is an element of length of the circuit C, and $\mathbf{n}\mathrm{d}a$ is an element of the area A enclosed by C. The negative sign reflects the further observation, made by Lenz, that the direction of the electromotive force is such as to give rise to a current which opposes the change of magnetic flux.

The integral relation (B.1.9) is more useful in differential form. This is derived by way of **Stokes' theorem**, which states that for any vector $\mathbf{E}$

$$\oint_C \mathbf{E}.\mathbf{dl} = \int_A (\nabla\times\mathbf{E}).\mathbf{n}\ \mathrm{d}a \tag{B.1.10}$$

where the vector differential operation $\nabla\times$ is known as 'taking the **curl**' of the succeeding vector (see Exercise B.1.3(a)). In Cartesian coordinates (see Exercise B.1.3(b))

$$\nabla\times\mathbf{E} \equiv \begin{vmatrix} \mathbf{u}_x & \mathbf{u}_y & \mathbf{u}_z \\ \dfrac{\partial}{\partial x} & \dfrac{\partial}{\partial y} & \dfrac{\partial}{\partial z} \\ E_x & E_y & E_z \end{vmatrix} \tag{B.1.11}$$

From Equations (B.1.9) and (B.1.10), **Faraday's law** for a differential element is

$$\nabla \times \mathbf{E} = -\frac{\partial \mathbf{B}}{\partial t} \quad , \qquad \text{(B.1.12)}$$

which will be referred to as the **third of Maxwell's equations.** It relates the positional properties of **E** to the time varying properties of **B**.

Exercises B.1.3

(a) The curl of a vector field is defined as having a component in a particular direction obtained by placing a small area normal to that direction, then, by letting the area become vanishingly small, the line integral around that area per unit area is the magnitude of the component of the curl. Use this definition to prove Stokes's theorem.

(b) Using the definition of curl in Exercise (a), find $\nabla \times \mathbf{E}$ when **E** is expressed in terms of its Cartesian coordinates E_x, E_y, and E_z.

(c) Starting with Equation (B.1.12), prove the second of Maxwell's equations. Note that the divergence of the curl of any vector is zero.

B.1.4 Time variation of electric field

In the case of a steady direct current, it follows from the definitions introduced in section B.1.2 that the closed line integral of the magnetic field **H** is equal to the current enclosed. (This is sometimes referred to as **Ampère's circuital law,** since it is based on his experiments.) If the current density is **J**′, then

$$\oint_C \mathbf{H}.\mathrm{d}\mathbf{l} = \int_A \mathbf{J}'.\mathbf{n}\mathrm{d}a \quad , \qquad \text{(B.1.13)}$$

and making use of Stokes' theorem

$$\nabla \times \mathbf{H} = \mathbf{J}' \quad . \qquad \text{(B.1.14)}$$

In attempting to extend this result to time-varying fields Maxwell found that it was necessary to postulate the existence of a 'displacement current' density $\partial \mathbf{D}/\partial t$ in addition to the conduction (or convection) current density **J**. The plausibility of this can be seen by considering the example of an alternating current source connected by two metal wires to a capacitor which has a perfect dielectric. Taking the line integral of the magnetic field around a closed line which passes once through the above simple circuit, this closed line can bound a surface which passes either completely through the wire or completely through the dielectric. When the surface passes through the wire the total current passing through the surface is the conduction current in the wire, which is equal to the line integral of the magnetic field. But if the surface is chosen such that it passes through the dielectric, since the latter is assumed perfect, there is no movement of charge through the surface. So if the definition of current is restricted to the movement of charge there will be an inconsistency in this particular case, since the value of the line integral of magnetic field is the same whether it bounds a surface passing through the wire or through the dielectric. However, by defining current as the sum of two parts, one which depends on the movement of charge and another which depends on the rate of change of electric flux density, the inconsistency is avoided. Indeed, in the present example it may be demonstrated (see Exercise B.1.4(c)) that the rate of change of electric flux in the dielectric, which is the displacement current, is equal to the charging current in the wire.

In general, therefore, the current density must be written as

$$\mathbf{J}' = \mathbf{J} + \frac{\partial \mathbf{D}}{\partial t} \quad . \tag{B.1.15}$$

and it follows from Equation (B.1.14) that the curl of the magnetic field is

$$\nabla \times \mathbf{H} = \mathbf{J} + \frac{\partial \mathbf{D}}{\partial t} \quad . \tag{B.1.16}$$

This will be referred to as the **fourth of Maxwell's equations.** The conduction current density may be given in terms of the bulk conductivity σ (siemens per metre) of the medium, thus

$$\mathbf{J} = \sigma \mathbf{E} \tag{B.1.17}$$

or in terms of the movement of charges as

$$\mathbf{J} = \rho \mathbf{v} \tag{B.1.18}$$

where ρ is the charge density and $\mathbf{v}$ is the velocity.

Exercises B.1.4

(a) Given that the magnetic field due to an infinitely long straight wire carrying a current I is $H = I/2\pi r$ in a circumferential direction at a distance r from the wire, show that the line integral of magnetic field around an arbitrary closed path lying in a plane perpendicular to the wire has the value I.

(b) The law of conservation of charge (**Franklin's law**) can be expressed as

$$\nabla . \mathbf{J} = -\,\partial\rho/\partial t \tag{B.1.19}$$

which states that the divergence of current density is equal to the rate of loss of charge. Using this, prove Maxwell's first equation by taking the divergence of the fourth.

(c) Find the electric flux density between the plates of an ideal, parallel-plate capacitor whose plates are charged to $+q$ and $-q$. (Use Gauss's law, and assume that the field is confined to parallel lines between the plates). Hence show that the total displacement current is equal to the charging current.

B.2 SOME CONSEQUENCES AND PARTICULAR FORMS OF MAXWELL'S EQUATIONS

Given the basic laws, embodied in Maxwell's equations, which govern the behaviour of electric and magnetic fields, certain important consequences follow. The first three to be considered here are: the superposition of solutions, Poynting's theorem for the flow of electromagnetic energy, and the conditions imposed on the fields at a boundary between dielectrics or at a perfectly conducting surface. Then follow three particular forms of Maxwell's equations: solutions that apply in source-free lossless media, solutions for sinusoidal time variations, and solutions in lossy media. Finally, three important general principles, concerning duality, reciprocity and uniqueness, will be derived.

B.2.1 Linearity. Superposition of Solutions

Summarizing the results of section B.1, Maxwell's equations in differential form are

$$\nabla.\mathbf{D} = \rho \tag{B.2.1}$$

$$\nabla.\mathbf{B} = 0 \tag{B.2.2}$$

$$\nabla \times \mathbf{E} = -\frac{\partial \mathbf{B}}{\partial t} \tag{B.2.3}$$

$$\nabla \times \mathbf{H} = \mathbf{J} + \frac{\partial \mathbf{D}}{\partial t} \tag{B.2.4}$$

in which

$$\mathbf{D} = \epsilon \mathbf{E} \tag{B.2.5}$$

and

$$\mathbf{B} = \mu \mathbf{H} \quad . \tag{B.2.6}$$

It should be noted that since equations (B.2.1) and (B.2.2) may be derived from equations (B.2.4) and (B.2.3), it is sufficient to consider equations (B.2.3) and (B.2.4) as Maxwell's equations.

The medium is said to be **linear** if the parameters ϵ and μ are indpendent of the fields **E** and **H**. In a linear medium, therefore, Maxwell's equations are linear equations, since differentiation with respect to either space or time variables is a linear operation. Hence separate solutions of Maxwell's equations may be superimposed to give a composite solution.

B.2.2 Poynting's Theorem. Flow of Electromagnetic Energy

Consider the equations

$$\nabla \times \mathbf{E} = -\frac{\partial \mathbf{B}}{\partial t} \tag{B.2.7}$$

and

$$\nabla \times \mathbf{H} = \mathbf{J} + \frac{\partial \mathbf{D}}{\partial t} \tag{B.2.8}$$

to apply to a linear medium which contains no free charges but is lossy, with conductivity σ. Then $\mathbf{J} = \sigma \mathbf{E}$.

From the vector identity

$$\nabla.(\mathbf{E} \times \mathbf{H}) = \mathbf{H}.(\nabla \times \mathbf{E}) - \mathbf{E}.(\nabla \times \mathbf{H}) \tag{B.2.9}$$

$$= -\mathbf{H}.\frac{\partial \mathbf{B}}{\partial t} - \mathbf{E}.\frac{\partial \mathbf{D}}{\partial t} - \mathbf{E}.\mathbf{J}$$

$$= -\frac{\partial}{\partial t}\left(\frac{\mu H^2}{2} + \frac{\epsilon E^2}{2}\right) - \sigma E^2 \tag{B.2.10}$$

Integrating over a closed volume V, whose surface is A, and applying the divergence theorem,

$$-\oint_A (\mathbf{E} \times \mathbf{H}).\mathbf{n}\ da = \int_V \left[\frac{\partial}{\partial t}\left(\frac{\mu H^2}{2} + \frac{\epsilon E^2}{2}\right) + \sigma E^2\right] dV \ . \tag{B.2.11}$$

The right-hand side of this equation is the rate of increase in stored magnetic energy $\mu H^2/2$ and stored electric energy $\epsilon H^2/2$ and the loss σE^2 over the whole volume V. The **Poynting vector**

$$\mathbf{S} = \mathbf{E} \times \mathbf{H} \tag{B.2.12}$$

may be interpreted as the instantaneous power flow, since the left hand side of equation (B.2.11) is the flux of **S** *into* the volume. The direction of **S** is the direction of power flux, which has the units of watts per square metre.

Exercises B.2.2

(a) Use the circuit result, that the electric energy stored in a capacitor is $\frac{1}{2}CV^2$, where C is the capacitance and V is the voltage, to demonstrate that the stored electric energy in a medium of permittivity ϵ is $\frac{1}{2}\epsilon E^2$ per unit volume.

(b) An infinitely long, straight wire of resistance R per unit length carries a steady current I. If the wire is circular in cross-section, show that the Poynting vector at the surface of the wire is directed radially inwards toward the centre of the wire, and that the total flux of power into the wire is equal to the resistive loss.

(c) Prove the vector identity of equation (B.2.9).

(d) Use Gauss's law (section B.1.1) to prove the **divergence theorem**, that

$$\int_V \nabla.\mathbf{D}\ \mathrm{d}V = \oint_A \mathbf{D}.\mathbf{n}\ \mathrm{d}a \quad .$$

B.2.3 Boundary Conditions

Maxwell's equations in integral form may be applied to vanishingly small regions whch straddle the boundary separating two electrically different media in order to obtain the relationship between the fields on either side of the boundary. The resulting **boundary conditions** are as follows; the two media being denoted by the subscripts 1 and 2, and the unit vector **n** which is normal to the boundary being assumed to be directed away from region 2 towards region 1.

The first of Maxwell's equations, which in integral form is Gauss's law, gives that the difference in the normal component of electric flux density across the boundary is equal to the surface density of charge ρ_s. In vector form this is

$$\mathbf{n}.(\mathbf{D}_1 - \mathbf{D}_2) = \rho_s \tag{B.2.13}$$

where ρ_s is in coulombs per square metre.

The second of Maxwell's equations, which in integral form states that the total magnetic flux out of a closed volume is zero, gives that the difference in the normal component of magnetic flux density across a boundary is zero; or, in other words, that the normal component of magnetic flux density is continuous across any boundary. In vector form

$$\mathbf{n}.(\mathbf{B}_1 - \mathbf{B}_2) = 0 \quad . \tag{B.2.14}$$

From Faraday's law, expressed as the line integral around a closed path of the electric field being equal to the negative rate of change of the magnetic flux enclosed, it may be shown that the tangential components of electric field on either side of the boundary are equal, so that

$$\mathbf{n} \times (\mathbf{E}_1 - \mathbf{E}_2) = 0 \quad . \qquad \text{(B.2.15)}$$

Finally, from Ampère's circuital law, which relates the closed line integral around a closed path of magnetic field to the total current enclosed, it follows that the tangential components of magnetic field on either side of the boundary are also equal, so that

$$\mathbf{n} \times (\mathbf{H}_1 - \mathbf{H}_2) = 0 \qquad \text{(B.2.16)}$$

The above four boundary conditions are based on the reasonable physical assumption that none of the electrical quantities involved in either region is infinite. However, the ideal case where one of the media is perfectly conducting (and therefore has infinite conductivity) is so often a realistic approximation in practice that the appropriate boundary conditions will be given here. It is clear from the analysis in section A.7 that fields cannot penetrate a perfect conductor, and so all fields must be zero within the metal. Therefore in equations (B.2,13-16) the field quantities with subscript 2 (assumed to refer to the metal region) are zero and the subscript 1 may therefore be omitted. Further, it was also shown in section A.7 for the penetration of electromagnetic waves into good conductors, that the current associated with the magnetic field flows in a thin layer at the surface of the metal. In the limit of infinite conductivity this layer becomes infinitesimal, so that the infinite current density must be treated as a surface current density $\mathbf{J}_s$. The application of Ampère's circuital law at the boundary of a perfect conductor will therefore incorporate this surface current in the boundary condition.

The boundary conditions appropriate to the fields just outside a perfectly conducting medium are therefore

$$\mathbf{n}.\mathbf{D} = \rho_s \qquad \text{(B.2.17)}$$

$$\mathbf{n}.\mathbf{B} = 0 \qquad \text{(B.2.18)}$$

$$\mathbf{n} \times \mathbf{E} = 0 \qquad \text{(B.2.19)}$$

$$\mathbf{n} \times \mathbf{H} = \mathbf{J}_s \qquad \text{(B.2.20)}$$

Exercises 1.7

(a) Deduce the boundary conditions given in equations (B.2.13–16).

(b) Show that if charge is introduced into a conducting medium, whose conductivity is σ, it will dissipate exponentially with time, having a time constant ϵ/σ. Where does the charge dissipate to?

B.2.4 Maxwell's Equations in Source-Free, Lossless Media

If net charge or current is introduced by some means into a region they will give rise to electric and magnetic fields in a manner already discussed. If such charges and currents are excluded from the region under consideration then it is said to be **source-free.** If, in addition, the region has zero conductivity the only currents will be displacement currents, and Maxwell's equations take on the simpler form

$$\nabla . \mathbf{D} = 0 \tag{B.2.21}$$

$$\nabla . \mathbf{B} = 0 \tag{B.2.22}$$

$$\nabla \times \mathbf{E} = -\partial \mathbf{B}/\partial t \tag{B.2.23}$$

$$\nabla \times \mathbf{H} = \partial \mathbf{D}/\partial t \tag{B.2.24}$$

B.2.5 Sinusoidal Time Variations. Average Power flow

A further simplification is obtained if it is assumed that the variation in time of the fields is purely sinusoidal. This implies that only one frequency is involved, and so such fields are often referred to as **monochromatic.** If more than one frequency is involved, as is usually the case in practice, the linearity of Maxwell's equations allows the complete solution to be obtained by the superposition of several single-frequency solutions.

If attention were to be confined to one component of, say, the electric field at a fixed point in space, this quantity is no different from a sinusoidally time-varying quantity in a circuit. It may therefore be represented by the complex quantity E_o, known as a **phasor,** such that the actual value at a particular time is given by

$$E(t) = \text{Re}\ [E_o e^{j\omega t}] \tag{B.2.25}$$

where Re indicates that the real part is taken. Then since E_o in general has a magnitude $|E_o|$ and phase ϕ,

$$E(t) = \text{Re}\ [|E_o| e^{j(\omega t+\phi)}] = |E_o| \cos(\omega t+\phi) \tag{B.2.26}$$

Thus sinusoidally time-varying quantities can be dealt with in a complex form which has no explicit indication of variation with time, on the understanding that the actual value at a particular time t is obtained by taking the real part of the product of the complex form and $e^{j\omega t}$. The complex form is much simpler to use, and is valid provided linear operations are involved.

A brief examination of equations (B.2.25-26) shows that differentation with respect to time of $E(t)$ has the complex form $j\omega E_o$. Thus the differential operator

$$\partial/\partial t \equiv j\omega \quad . \tag{B.2.27}$$

The same considerations apply to a vector quantity, where, in rectangular coordinates,

$$\mathbf{E} = \mathbf{u}_x E_x + \mathbf{u}_y E_y + \mathbf{u}_z E_z$$

and each component depends on both space and time. The x-component has the complete functional representation

$$E_x(x,y,z,t) = \text{Re}\ [E_x(x,y,z)\, e^{j\omega t}]$$

in which $E_x(x,y,z)$ is complex. Then the vector field

$$\mathbf{E}(x,y,z,t) = \text{Re}\ [\mathbf{E}(x,y,z)\, e^{j\omega t}] \tag{B.2.28}$$

is obtained from the complex (or phasor) vector $\mathbf{E}(x,y,z)$. Conventionally the complex-vector form for $\mathbf{E}$ will be used, the operation shown by equation (B.2.28) being implied.

Maxwell's equations for a source-free, lossless medium, with sinusoidal time variations therefore have the form

$$\nabla.\mathbf{D} = 0 \tag{B.2.29}$$

$$\nabla.\mathbf{B} = 0 \tag{B.2.30}$$

$$\nabla \times \mathbf{E} = -j\omega\mathbf{B} \tag{B.2.31}$$

$$\nabla \times \mathbf{H} = j\omega\mathbf{D} \quad . \tag{B.2.32}$$

The average power flow is given by the **average Poynting vector**

$$\mathbf{S}_{\text{av}} = \tfrac{1}{2}\ \text{Re}\ [\mathbf{E} \times \mathbf{H}^*] \quad , \tag{B.2.33}$$

by analogy with the circuit-theory construction for obtaining average power per cycle when using complex forms (the asterisk denotes the complex conjugate).

Exercise B.2.5

(a) Derive the **wave equation**

$$\nabla^2 \mathbf{E} + k^2 \mathbf{E} = 0$$

or

$$\nabla^2 \mathbf{H} + k^2 \mathbf{H} = 0$$

for sinusoidal fields in a source-free, uniform, isotropic and loss-less medium where $k^2 = \omega^2 \mu \epsilon$. Take the curl of one of Maxwell's curl equations and then make use of the identity in equation (2.13).

(b) Demonstrate the validity of equation (B.2.33).

B.2.6 Lossy Media. Complex Permittivity Representation

In a perfectly lossless medium whose permittivity and permeability are scalar functions of position, Maxwell's equations are

$$\nabla \times \mathbf{E} = -j\omega\mu\mathbf{H} \tag{B.2.34}$$

and

$$\nabla \times \mathbf{H} = j\omega\epsilon\mathbf{E} \tag{B.2.35}$$

If the medium has finite conductivity σ, Maxwell's equations must include the conduction current so they become

$$\nabla \times \mathbf{E} = -j\omega\mu\mathbf{H} \tag{B.2.36}$$

and

$$\nabla \times \mathbf{H} = \sigma\mathbf{E} + j\omega\epsilon\mathbf{E} \quad . \tag{B.2.37}$$

Comparing equations (B.2.34-35) with (B.2.36-37), it is apparent that a lossy medium can be treated as though it were lossless by employing the complex permittivity

$$\epsilon_c = \epsilon + \frac{\sigma}{j\omega} \tag{B.2.38}$$

in the equations for a lossless medium. Thus the simpler case of a lossless medium can be analysed and the results applied directly to a lossy medium by the substitution for ϵ of the complex ϵ_c of equation (B.2.38).

B.2.7 Duality of Solutions

Making the following substitutions

$$\mathbf{E} = Z\mathbf{H}' \tag{B.2.39}$$

and

$$\mathbf{H} = -\frac{\mathbf{E}'}{Z} \ , \tag{B.2.40}$$

where

$$Z = \sqrt{\frac{\mu}{\epsilon}} \tag{B.2.41}$$

in the equations

$$\nabla \times \mathbf{E} = -j\omega\mu\mathbf{H} \tag{B.2.42}$$

$$\nabla \times \mathbf{H} = j\omega\epsilon\mathbf{E} \tag{B.2.43}$$

shows that $\mathbf{E}'$ and $\mathbf{H}'$ are also solutions of Maxwell's equations as applied to sinusoidal fields in source-free, lossless media. This means that if one solution for the fields **E** and **H** in such a situation has been found, then a second solution $\mathbf{E}'$ and $\mathbf{H}'$ can be obtained by using the relations (B.2.39-40).

B.2.8 Reciprocity Theorems

Consider two elemental current sources, $\mathbf{I}_1$ at $\mathbf{r} = \mathbf{r}_1$ and $\mathbf{I}_2$ at $\mathbf{r} = \mathbf{r}_2$, where **r** is the position vector measured from an arbitrary origin O. The two current sources produce respectively the two field distributions ($\mathbf{E}_1$, $\mathbf{H}_1$) and ($\mathbf{E}_2$, $\mathbf{H}_2$). If both sources oscillate at the same frequency ω, Maxwell's equations for the two sets of fields and their sources are

$$\nabla \times \mathbf{E}_1 = -j\omega\mathbf{B}_1 \tag{B.2.44}$$

$$\nabla \times \mathbf{H}_1 = j\omega\mathbf{D}_1 + \mathbf{I}_1\,\delta(\mathbf{r} - \mathbf{r}_1) \tag{B.2.45}$$

and

$$\nabla \times \mathbf{E}_2 = -j\omega\mathbf{B}_2 \tag{B.2.46}$$

$$\nabla \times \mathbf{H}_2 = j\omega\mathbf{D}_2 + \mathbf{I}_2\,\delta(\mathbf{r} - \mathbf{r}_2) \ . \tag{B.2.47}$$

The above equations refer to the general field point $\mathbf{r}$; the three-dimensional delta function $\delta(\mathbf{r}-\mathbf{r}')$ is zero everywhere except at $\mathbf{r}=\mathbf{r}'$ where the integral

$$\int_V f(\mathbf{r})\,\delta(\mathbf{r}-\mathbf{r}')\,\mathrm{d}V = f(\mathbf{r}') \tag{B.2.48}$$

over a volume V which contains the point $\mathbf{r}=\mathbf{r}'$.

The divergence

$$\nabla.(\mathbf{E}_1\times\mathbf{H}_2-\mathbf{E}_2\times\mathbf{H}_1)=\mathbf{H}_2.(\nabla\times\mathbf{E}_1)-\mathbf{E}_1.(\nabla\times\mathbf{H}_2)-\mathbf{H}_1.(\nabla\times\mathbf{E}_2)$$
$$+\mathbf{E}_2.(\nabla\times\mathbf{H}_1)$$

Substituting from equations (B.2.44-47) and using the relations $\mathbf{B}=\mu\mathbf{H}$ and $\mathbf{D}=\epsilon\mathbf{E}$ yields

$$\nabla.(\mathbf{E}_1\times\mathbf{H}_2-\mathbf{E}_2\times\mathbf{H}_1)=-\,\mathbf{E}_1.\mathbf{I}_2\,\delta(\mathbf{r}-\mathbf{r}_2)+\mathbf{E}_2.\mathbf{I}_1\,\delta(\mathbf{r}-\mathbf{r}_1)$$

Taking the volume integral, and using the divergence theorem (Exercise B.2.2(d)) together with equation (B.2.48)

$$\oint_A(\mathbf{E}_1\times\mathbf{H}_2-\mathbf{E}_2\times\mathbf{H}_1).\mathbf{n}\,\mathrm{d}a=-\mathbf{E}_1(\mathbf{r}_2).\mathbf{I}_2+\mathbf{E}_2(\mathbf{r}_1).\mathbf{I}_1 \tag{B.2.49}$$

where the volume whose surface is A encloses both sources.

It is known (see section 3.5.3, for example) that the magnitudes of fields such as $\mathbf{E}_1$, $\mathbf{H}_1$, etc., fall off inversely as the distance from the source, provided this is large. Hence if the sources $\mathbf{I}_1$ and $\mathbf{I}_2$ exist in an unbounded medium the surface integral in equation (B.2.49) can be taken at infinity. Then, provided the medium can be assumed to be slightly lossy, the surface integral is zero and

$$\mathbf{E}_1(\mathbf{r}_2).\mathbf{I}_2=\mathbf{E}_2(\mathbf{r}_1).\mathbf{I}_1 \quad . \tag{B.2.50}$$

Equation (B.2.50) states that the component of electric field (due to the first source) at the second source and parallel to it equals the component (due to the second source) at the first and parallel to it. Putting $\mathbf{I}_1=\mathbf{I}_2$ in equation (B.2.50) leads to the practical statement of reciprocity, that if the source and receiver are interchanged the received signal remains the same.

It should be noted that the surface integral in equation (B.2.49) will still be zero if part or all of the surface, not necessarily at infinity, is composed of conducting material. This is obviously true for a perfect conductor since the tangential electric field must always vanish at the surface. But it is also true for an imperfect conductor provided the surface A lies well below the skin depth, in which case all field components are essentially zero (see section A.6).

An alternative form of reciprocity theorem is obtained by taking the volume of integration, whose bounding surface is A, to exclude the sources which produce the fields. Then instead of equation (B.2.49) we have that

$$\oint_A (\mathbf{E}_1 \times \mathbf{H}_2 - \mathbf{E}_2 \times \mathbf{H}_1).\mathbf{n}\ da = 0 \quad . \tag{B.2.51}$$

This is known as the **Lorentz reciprocity theorem**, and is useful when only fields (rather than sources) are specified.

B.2.9 Uniqueness Theorem

Consider a source-free volume V completely enclosed by the surface A. Suppose that there are two solutions ($\mathbf{E}_1$, $\mathbf{H}_1$) and ($\mathbf{E}_2$, $\mathbf{H}_2$) which both satisfy Maxwell's equation within V and on its surface A. Then by the principle of superposition (section B.2.1) the difference fields

$$\mathbf{E}_1 - \mathbf{E}_2 \text{ and } \mathbf{H}_1 - \mathbf{H}_2 \tag{B.2.52}$$

will also satisfy Maxwell's equations.

Now apply the divergence theorem (as in the proof of Poynting's theorem, section B.2.2) with $\mathbf{n}$ the outward normal to the surface,

$$-\oint_A (\mathbf{E}_1 - \mathbf{E}_2) \times (\mathbf{H}_1 - \mathbf{H}_2).\mathbf{n}\ da$$

$$= \int_V \left[\frac{\partial}{\partial t}\left(\frac{\mu|\mathbf{H}_1 - \mathbf{H}_2|^2}{2} + \frac{\epsilon|\mathbf{E}_1 - \mathbf{E}_2|^2}{2}\right) + \sigma|\mathbf{E}_1 - \mathbf{E}_2|^2\right] dV \quad . \tag{B.2.53}$$

The normal components of the difference fields ($\mathbf{E}_1 - \mathbf{E}_2$) and ($\mathbf{H}_1 - \mathbf{H}_2$) do not enter into the integral on the left hand side. Hence if *either* tangential **E** *or* tangential **H** is specified uniquely at all points on A, then the integral on the left hand side is zero. For the right-hand side to be zero, both ($\mathbf{E}_1 - \mathbf{E}_2$) and ($\mathbf{H}_1 - \mathbf{H}_2$) must be identically zero throughout the entire volume. Hence there is one, and only one, solution throughout V (that is, the solution is **unique** within V) if either $\mathbf{E}_{\tan}$ or $\mathbf{H}_{\tan}$ is specified over the entire surface A.

References

Abramowitz, M. and Stegun, I. A. (1964) *Handbook of mathematical functions with formulas, graphs, and mathematical tables.* National Bureau of Standards, U.S. Government Printing Office, Washington, D. C.

Booker, H. G. (1947) The elements of wave propagation using the impedance concept. *Journal IEE,* **94,** PtIII, 171-198.

Booker, H. G. and Clemmow, P. C. (1950) The concept of an angular spectrum of plane waves, and its relation to that of polar diagram and aperture distribution. *Proc. IEE,* **97,** PtIII, 11-17.

Born, M. and Wolf, E. (1970) *Principles of optics* 4th Edn. Pergamon, Oxford.

Brown, J. (1958a) A theoretical analysis of some errors in aerial measurements *Proc. IEE,* **105,** Pt.C, 343-351.

Brown, J. (1958b) A generalized form of the aerial reciprocity theorem *Proc. IEE,* **105,** Pt.C, 472-475.

Brown, J. and Glazier, E. V. D. (1974) *Telecomunications.* 2nd Edn: Chapman and Hall, London.

Chappas, C. C. (1974) The calculation of aperture radiation patterns. PhD thesis, University of London.

Chu, T. S. (1965) An approximate generalization of the Friis transmission formula, *Proc. IEEE,* **53**(3), 296-297.

Clarke, R. H. and Hendry, G. O. (1964) Prediction and measurement of the coherent and incoherent power reflected from a rough surface. *Trans. IEEE,* **AP-12**, (3), 353-363.

Clarricoats, P. J. B. and Poulton, G. T. (1967) High-efficiency microwave reflector antennas – a review. *Proc. IEEE,* **65**(10), 1470-1504.

Clemmow, P. C. (1966) *The plane wave spectrum representation of electromagnetic fields.* Pergamon, Oxford.

Crawford, A. B., Hogg, D. C. and Hunt, L. E. (1961) Project ECHO: a horn-reflector antenna for space communication. *Bell System Tech. J.,* **40**(4), 1095-1116. Also in Love (1976).

Debye, P. (1909) Das Verhalten von Lichtwellen in der Nähe eines Brennpunktes oder einer Brennlinie. *Ann. Phys.*, **30**, 755-776.

Dennery, P. and Krzywicki, A. (1967) *Mathematics for physicists* Harper and Row, New York.

Friis, H. T. (1946) A note on a simple transmission formula. *Proc. IRE*, **34**(5), 254-256.

Gradshteyn, I. S. and Ryzhik, I. M. (1965) *Tables of integrals, series, and products* 4th Edn. Academic Press, New York.

Harrington, R. F. (1959) On scattering by large conducting bodies. *Trans. IRE*, **AP-7**(2), 150-153.

Harrington, R. F. (1961) *Time-harmonic electromagnetic fields*. McGraw-Hill, New York.

Jahnke, E. and Emde, F. (1945) *Tables of functions with formulae and curves* 4th Edn. Dover, New York.

Johnson, R. C., Ecker, H. A. and Hollis, J. S. (1973) Determination of far-field antenna patterns from near-field measurements. *Proc. IEEE*, **61**(12), 1668-1694.

Jordan, E. C. and Balmain, K. G. (1968) *Electromagnetic waves and radiating systems* 2nd Edn. Prentice-Hall, New Jersey.

Joy, E. B., Leach, W. M., Rodrigue, G. P. and Paris, D. T. (1978) Applications of probe-compensated near-field measurements. *Trans IEEE*, **AP-26**(3), 379-389.

Keller, J. B. (1962) Geometrical theory of diffraction. *Journal Optical Soc. Am.*, **52**(2), 116-130.

Love, A. W. (ed.) (1978) *Electromagnetic horn antennas.* IEEE Press, New York.

Love, A. W. (ed.) (1978) *Reflector antennas.* IEEE Press, New York.

Maxwell, J. Clerk (1873) *A treatise on electricity and magnetism.* Clarendon Press, Oxford.

Ohm, E. A. (1961) Project ECHO: receiving system *Bell System Tech. J.* **40**(4), 1065-1094.

Paris, D. T., Leach, W. M. and Joy, E. B. (1978) Basic theory of probe-compensated near-field measurements. *Trans. IEEE*, **AP-26**(3), 373-379.

Penzias, A. A. and Wilson, R. W. (1965) A measurement of excess antenna temperature at 4080 Mc/s. *Astrophysical Journal.* **142**(1), 419-421.

Plonsey, R. and Collin, R. E. (1961) *Principles and applications of electromagnetic fields.* McGraw-Hill, New York.

Ramo, S., Whinnery, J. R., and Van Duzer, T. (1965) *Fields and waves in communication electronics,* 2nd Edn. Wiley, New York.

Ratcliffe, J. A. (1956) Some aspects of diffraction theory and their application to the ionosphere. *Reports on Progress in Physics,* **19**, 188-267.

Rayleigh, Lord (1896) *The theory of sound.* Part 2. 2nd Edn. Macmillan, London. (1945) Dover, New York.

Rhodes, D. R. (1974) *Synthesis of planar antenna sources.* Clarendon, Oxford.

Rice, S. O. (1951) Reflection of electromagnetic waves from slightly rough surfaces. *Communications on Pure and Applied Mathematics,* **4**(2/3), 351-378.

Ross, R. A. (1966) Radar cross section of rectangular flat plates as a function of aspect angle. *Trans IEEE,* **AP-14**(3), 329-335.

Rudge, A. W. and Adatia, N. A. (1978) Offset-parabolic-reflector antennas: A review. *Proc. IEEE,* **66**(12), 1592-1618.

Schelkunoff, S. A. (1943) *Electromagnetic waves.* Van Nostrand, New Jersey.

Schensted, C. E. (1955) Electromagnetic and acoustic scattering by a semi-infinite body of revolution. *Jour. Appl. Phys.,* **26**(3), 306-308.

Silver, S. (1949) *Microwave antenna theory and design.* McGraw-Hill, New York.

Stratton, J. A. (1941) *Electromagnetic theory* McGraw-Hill, New York.

Weyl, H. (1919) Ausbreitung elektromagnetischer Wellen über einen ebenen Leiter *Ann. Phys.* **60**, 481-500.

Whittaker, E. T. (1902) On the partial differential equations of mathematical physics. *Math. Ann.* **57**, 333-355.

Woodward, P. M. and Lawson, J. D. (1948) The theoretical precision with which an arbitrary radiation pattern may be obtained from a source of finite size. *Journal IEE,* **95**, PartIII, 363-370.

Index

B

C

G

L

M

N

O

P

X

Y

Z